Architectural Engineering and Design Management

TEACHING AND LEARNING BUILDING DESIGN AND CONSTRUCTION

EDITOR: Dino Bouchlaghem

GUEST EDITORS: David Dowdle and Vian Ahmed

Architectural Engineering and Design Management: an international journal

Bridging the gap between architecture and engineering practice

Influential and far-reaching, *Architectural Engineering and Design Management* provides a unique forum for the dissemination of academic and practical developments related to architectural engineering and building design management. This new international, peer-reviewed journal details the latest cutting-edge research and innovation within the field, spearheading improved efficiency in the construction industry. Informative and accessible, this publication analyses and discusses the integration of the main stages within the process of design and construction and multidisciplinary collaborative working between the different professionals involved.

Ideal for practitioners and academics alike, *Architectural Engineering and Design Management* examines specific topics on architectural technology, engineering design, building performance and building design management to highlight the interfaces between them and bridge the gap between architectural abstraction and engineering practice. Coverage includes:

- Integration of architectural and engineering design
- Integration of building design and construction
- Building design management; planning and co-ordination, information and knowledge management, value engineering and value management
- Collaborative working and collaborative visualisation in building design
- Architectural technology
- Sustainable architecture
- Building thermal, aural, visual and structural performance
- Education and architectural engineering

ISSN 1745-2007

Visit www.earthscan.co.uk for further information about this journal

JOURNAL SUBSCRIPTION INFORMATION

Institutional subscriptions are US$325, UK£195 (airmail extra); personal subscriptions are US$160, UK£95 (airmail extra).

First published by Earthscan in the UK and USA in 2006

This edition published 2013 by Earthscan

For a full list of publications please contact:

Earthscan
2 Park Square, Milton Park, Abingdon, Oxon OX14 4RN
Simultaneously published in the USA and Canada by Earthscan
711 Third Avenue, New York, NY 10017

Earthscan is an imprint of the Taylor & Francis Group, an informa business

Contents

Contents

Editorial

David Dowdle and Vian Ahmed (Guest Editors)

The editors of this special issue are keen proponents of innovation in teaching and learning. This interest is influenced by the familiar occurrences of increasing class size, loss of laboratory space, widening participation, worsening attendance, etc., which frequently lead to retention and progression problems. Academics are asking how they can support students whilst reducing the delivery and assessment burdens imposed by the changing student profile. This has been a subject of debate in many formal and informal discussions for a number of years. If anything, this is a reflection of the care of the academic community for future generations of industry practitioners. Support from government initiatives, funding bodies and academic institutes has been reflected by the quantity and quality of research undertaken to embed innovative teaching methods within higher education. This is witnessed by the number of publications received to enable this issue.

The special issue covers many of the named themes suggested in the original call, plus papers that fell within the overall 'teaching and learning in the built environment' remit and we feel this has resulted in a well balanced publication.

The ability of new architects, and new practitioners in general, to 'hit the ground running' when they start their careers is a problem oft quoted by the professions and industry alike when commenting on the qualities of graduates today. The paper by Wood identifies the separation of construction from design in architecture schools as one of the root causes of this malaise and offers a strategy to help 'bridge the widening divide between academia and architectural practice'.

Tucker and Rollo report on a problem facing many academics – how to deal with larger student numbers with the same staff resource. In the scenario discussed, rather than spreading themselves thinly over many individual projects, lecturers initiated group projects that allowed them to offer in-depth review sessions on a smaller number of projects. Assessment of individual skills and competences and the formation of groups are addressed. Findings conclude that the input from tutors in helping develop project groups is extensive and the notion that group projects demand less teaching resource is based on a false economy.

Several papers have 'e-learning' as their theme. David Dowdle's presents the case for introducing e-learning into construction education. A detailed literature review examines the activities required to achieve 'effective learning', with the key activity – 'student-centred learning' – being ideally suited to e-learning. The paper then compares the benefits of e-learning to more traditional methods before introducing a 'guidance/audit' tool aimed at assisting lecturers in the creation of new e-learning content or the assessment of existing content, all with the aim of ensuring user engagement and the creation of effective learning.

David Lowe's paper confirms the benefits to students and employers of an industrial placement are long established. Students check employers' ability to provide them with post-graduation experience and long-term careers, whilst employers use the opportunity to assess the capability of students prior to offering employment. An eight year study into the ability of construction organisations to provide appropriate learning environments during work experience is analysed. Findings suggest that, whilst they are getting many things right, organisations need to enhance their student support mechanisms if they wish to attract and retain potential graduate employees.

The teaching–research nexus, i.e. the interdependence between the research activity of staff and the learning experienced by their students, has long been a central feature of UK higher education. In his paper, Stephen Emmitt seeks to investigate the theoretical and practical links between teaching and research in a teaching-led UK university. A detailed literature review coupled with semi-structured interviews and reflections on the changes made to an architectural programme over a four year period contribute to a robust analysis of the issues and leads to several suggestions for real improvement.

Chinyio and Morton's paper involves research into e-learning delivered via the 'Moodle' VLE. Guidance is offered as to the enablers and inhibitors of e-learning based on their experiences as both tutees and tutors. Valuable insights from the learner's viewpoint and that of the developer/facilitator become evident. The paper focuses on the tools and facilities available rather than on content development. As tutees, they reflect on the tools that encourage and facilitate group discussion and, as tutors, they survey user perceptions of the interactive learning support tools available within the VLE.

The third paper with an e-learning theme examines the 'bigger picture' of e-learning development – sharing and re-use. Ahmed, Shaik and Aouad describe the relatively new technology of the 'Semantic Web', which is addressed by 'the extension of the current World Wide Web' that can be employed to provide a 'platform independent' common framework that will allow e-learning to be shared and reused across application, enterprise, and community boundaries.

Simulations are becoming popular in higher education as a means to introduce learners to 'real life' work scenarios where health and safety concerns and lack of resources would, for example, block site visits or laboratory work. Gribble *et al* report on a study of more than 250 undergraduates into the effectiveness of one such simulation. The study examines how students interacted with the simulation as a learning tool and also how teamworking and communication skills were developed.

When 'teaching and learning' is coupled with 'innovation' many academics and learning technologists look to the latest technologies such as e-learning or VLEs. They forget that one definition of innovation is the refinement or reapplication of existing technology. Hoxley and Rowsell examine how videos can be used more effectively in lectures. Findings suggest that students are receptive to their use but need encouragement to engage with the learning offered. To this end, the authors propose the use of a tightly focused interactive quiz as an 'orienting activity' to promote active learning.

Kumaraswamy *et al* report on two web-based tools that target the teaching/learning/training opportunities of university students and small and medium sized contractors. Constraints imposed on undergraduate teaching via reduced opportunities for site visits are being circumvented using 'virtual site visits' whilst specialist topics are being addressed using a computer-aided teaching and learning package. The second tool allows contractors to communicate seamlessly via information management platforms that provide project-specific library tools, knowledge management tools and training tools, all aimed at improving the performance of those projects.

Traditional distance-learning programmes are often updated by the latest technologies in attempts to reduce distribution costs and improve quality. The introduction of CD-ROM based content delivery, for example, has quickly been superseded or enhanced by web-based support via VLEs. The paper by Allan *et al* reviews a technological update to an existing course using a 'blended learning' approach – where the 'original investment' of paper-based resource was protected whilst providing students with an enhanced learning experience.

Knowledge management has been a popular management initiative for many years and, in the final paper, David Boyd reports on an innovative attempt to link knowledge management to higher education via a MSc. course. The knowledge-centric approach used enables learners to deal with incomplete knowledge – a growing problem in the industry. The aim is to create practitioners who are adequate with incomplete knowledge, can engage with the 'unfamiliar' as well as the 'familiar' and can 'hit the ground running' which, as already mentioned, is something employers are increasingly looking for.

As Guest Editors, we would like to thank Professor Dino Bouchlaghem for giving us the opportunity to engage and be in touch with the academic community from many parts of the world. We would also like to thank all referees for their contribution and valuable comments and, last but not least, all authors who made the publication of this issue possible, as well as those who expressed their interest in contributing.

We sincerely hope that you find this special issue both interesting and informative.

ARTICLE

Demystifying Construction

Technology in Architectural Education

Antony Wood

Abstract

Construction is often seen, among both students and staff at schools of architecture, as a distinct subject taught separately from design. It is often delivered by lectures, divorced from design studio and relying heavily on published sources. Even when it is integrated into design studio, it is often conducted as an add-on project, a separate exercise set after the design is 'finished'. This is, in the author's opinion, a flawed train of thought that is largely accountable for the increasing ill-preparation of architectural graduates for the realities of working practice and the corresponding frustration of the profession and industry who receive them. This paper presents an alternative approach to the teaching of construction in schools of architecture. Based on a model successfully implemented at the University of Nottingham, it explores the teaching methodology, learning outcomes, methods of assessment and possible implications for both schools of architecture and the profession as a whole. In doing this, it suggests a strategy to help bridge the widening divide between academia and architectural practice.

■ *Keywords* – Architecture; construction; practice; profession; teaching

INTRODUCTION: THE PROBLEM

> *Architectural education is in great danger, and has been for the past 20 years or so, of being hijacked by those whose real interests are words rather than buildings. (Peter Cook, in Chadwick, 2004: 6)*

It often takes several years of working in architectural practice after graduation until students of architecture arrive at the realization that the assembly of materials at 1:5 scale involves the same design process, awareness and rationale as the assembly of form and space at 1:500. This is the realization that the only difference between conceptual design and construction design is one of scale. The irony with this realization is that, on many occasions during their architectural education, the same students would have undoubtedly heard that very fact from tutors – a mantra to be chanted during construction lectures: 'detailing is no different to design'. But although most students can understand this as an abstract notion, many struggle to put it into practice. When the time comes to produce the 'construction drawings' to accompany the latest design project, most reach for the latest edition of the *Architect's Journal Working Details* (Dawson, 2003) and choose the slickest detail they think they can justify for their building. These students have simply learnt to imitate construction details as opposed to being able to apply construction principles.

> *Students who have learned imitation subjects have been involved in a certain process that has enabled them to acquire factual knowledge which is useful in a very limited range of situations. Much of what they have learned has no personal relevance to them (except as a form of gaining qualifications) or any connection with the real world it is supposed to explain. (Ramsden, 1992)*

So why is this? Why are architectural students able to make the abstract link between the processes of conceptual and constructional design, but are unable to put it, literally, into practice? This paper – and the approach to construction teaching it contains – is based largely on the premise that this is because tutors explain the process, but very few actually demonstrate it. There is a lack of demonstration that those design discussions about the massing of a building, the way occupants move through the spaces or the visual implications of the junctions, involve the same thought processes that should be involved in considering the termination of a roof edge, how that edge might conceal a gutter, and how the constituent parts of the assembly act in symbiosis to achieve the function.

> *All knowledge involves the use of symbols and the making of judgements in ways that cannot be expressed in words and can only be learnt in a tradition. Accordingly, the business of acquiring knowledge of any form is therefore to a greater or lesser extent something that cannot be done simply by solitary study of the symbolic expressions of knowledge. (Hirst, 1975)*

So students of architecture fail to grasp the notion that construction can be tackled from first principles, from common sense, from a blank sheet of paper and an appreciation of the result to be achieved and the palette at one's disposal. Additionally, construction lectures in many schools of architecture tend to concentrate on material properties or building elements in absentia from design, whereas construction studio asks for a product that requires appreciation of a link that is missing. So, construction as a subject becomes mystified; a skill deemed by students to be the special domain of architects with years of experience. Construction becomes detached from the design process; it becomes a separate subject, an addendum to the creation.

> *The form and conduct of the typical design studio today encapsulates architecture schools' entrenched isolation. Once considered an exemplar of project teaching, the studio is increasingly distanced from the real world it is intended to simulate. Technical knowledge is generally restricted to separate lectures on construction and materials. (Abel, 1998)*

This, of course, is not a positive situation. Consideration of technology – structure, environmental response, construction – has to be an integral part of the design process from concept through to completion. The historical development of the role of construction in architectural education has been well documented (e.g. Carpenter, 1997). Conclusions tend towards the belief that construction cannot be learnt purely from books. It cannot be understood by the assimilation of information. It must be investigated, attempted, engaged with. It is only in the act of attempting a solution that the intricacies of the problem can be fully appreciated. This paper is concerned ultimately with the integration of technology, and construction specifically, in design. However, especially with increasing student numbers, it is not possible to reject the lecture format as a teaching medium. Rather, the following methodology explains how, in the past few academic sessions of third-year construction lectures at the University of Nottingham, both the design studio and the construction site have been brought into the lecture theatre in an attempt to demystify construction, to make it relevant to the students, to demonstrate to them the essential link to the design process.

LECTURE COURSE METHODOLOGY: LEARNING BY DOING

The methodology in the teaching approach (summarized below) is based on each lecture being centred on an exercise involving a differing building system or set of materials that the student must engage with. The intention is that, in learning by doing rather than just by listening or reading, the student will appreciate the rationale behind the assembly of elements and gain confidence to approach the detailed design of their own studio projects from first principles rather than borrowed sources. The methodology consists of seven elements:

- background – generic information on material/element issued
- sample – physical handling of material/element by students
- case study – images of material/element implementation on site
- exercise – students undertake task, produce detailed drawings

- marking – students are assessed on the value of their solutions
- feedback – author's solution is presented and analysed in detail
- reinforcement – case study construction slides are again shown.

STAGE 1: INTRODUCTION OF MATERIALS AND BUILDING ELEMENTS

The material or building element for that week's lecture and exercise is first introduced in the form of a case study of a building with which the author has been involved (and which, obviously, uses that material). Rather than the lecture programme being presented as a series of bland construction materials e.g. brick/block, composite cladding, standing seam roofing, planar glazing, etc., it is presented as a series of building types e.g. the piece of retail urban infill, the 'glossy pages' museum, the high-profile transport terminal, etc.

Information on the generic properties of the material and its use in buildings (i.e. the background information) is issued as a handout together with a sample of the material (plus items from the associated assembly e.g. fixing bracket) which is passed around the group. It is vital for the students to have an opportunity to experience the material – to touch it, test it and understand it. The case study is presented in the form of slides; first, as an object in the realization of the design (the brief, client requirements, design solution, etc.), then with increasing focus on the usage of the particular material/element i.e. the construction sequence, and the implications and peculiarities of that particular site operation. It is important that the material system is supported by images of the actual construction, to enforce the essential link between the drawings and the detail. In this way, the construction site is brought into the lecture theatre. The presentation concludes with details that are the focus of that week's exercise – both the images of its exposed construction on site and the corresponding detailed drawings.

STAGE 2: THE EXERCISE

The week's exercise is next introduced (Figure 1). Each exercise assumes a particular 'real' scenario in an architectural office and is based on a task undertaken by the author at some stage in his practice career. The student assumes the role of an architectural assistant working in the practice and is given various pieces of information, in the form of attached drawings (Figure 2) or dimensional information passed on from a hypothetical technical representative, to enable him/her to carry out the exercise.

The objectives of the exercises change and increase in complexity each week. Early exercises are concerned primarily with appreciating that materials come with their own set of rules for usage (modular, structural, aesthetic, etc.) and that these requirements should be respected (or, at least, understood to enable conscious manipulation). Later exercises tend to focus directly on specific scenarios and the in-depth intricacies of the corresponding details.

Practically, the student completes the exercise in the day(s) following the lecture (Figure 3) and submits it for marking. Marking of the exercise is essential due to student attitudes (i.e. students only take an educational activity seriously if it is assessed – see 'Resourcing Implications' below) and forms an important part of the feedback to students that follows.

STAGE 3: FEEDBACK

The author's solution to the exercise is presented in depth at the next lecture, building up the rationale informing each decision in the assembly of the detail – functional, practical and aesthetic. The exercise-marking structure is explained, so that students understand why each mark is allocated. In attaching marks to fundamental decisions, it highlights the importance of the considerations in the detail, reinforcing the idea that each line in the drawing represents an element that is the result of a conscious, considered decision.

Once the solution to the exercise has been discussed and understood, the issues are then reinforced by presenting the same case study slides that were presented before the exercise was undertaken. Whereas before the exercise they were viewed by the students as individual images somehow contributing to the overall assembly, now each element is understood in terms of the role it fulfils and its relationship to the constituent parts around it (Figures 4 and 5). Similar details on other buildings are also shown

"DEMYSTIFYING CONSTRUCTION" Lecture Series

Exercise 3: Forest Shopping Experience, Nottingham – the Retail Mall
(From Lecture 6: Standing Seam Roofing)

Now that you are getting into the retail store rationalisation for Aldi, and looking forward to extending the same detailing disciplines to the whole building (!!?!), the Director pulls you off the project to assist with pressures elsewhere in the office. Another retail project is about to start on site and, unfortunately, sufficient resources have not been put into the working drawing packages. The contractor is currently screaming as the steelwork is about to arrive on site, but he has not enough information on the roof details, beyond the written specification. The Director knows that there could be a serious liability case to answer to if, as architects, you don't keep to the 'information release schedule', and thus asks you to drop everything you're working on and help out..........

The main problem is the external canopy to the shopping mall entrance, which is envisaged as a curving roof over a huge steel 'tree' structure, very much in keeping with the 'Robin Hood / Forest' shopping experience theme (sic). The structural engineer sent confirmation of the steel truss structure (composed of 90mm circular hollow sections) several months ago and this was worked up into a 1:100 sketch section which was faxed to site (see drawing *Sk616* attached). As it stands, this is the only drawing the contractor has to describe the construction of the canopy. The Director tells you to check the specification, as this should tell you all you need to know (!).

The specification section on the canopy roof reads as a *"65mm Key Bemo aluminium standing seam roofing top layer, supported by proprietary halter fixings off a 32mm profiled (trapezoidal) steel decking liner panel supported directly from the primary steel truss via 225mm Z-purlins on cleats (purlin zone = 250mm). The zone between Key Bemo roofing and top of liner panel contains 100mm rockwool fibreglass insulation compressed by the construction to approximately 83mm and a 2mm thick Butyl rubber sealant sheet acting as a vapour control layer lain on top of the liner. Curving moisture-resistant plasterboard (multi-board) fixed to the undersides of the purlins forms a ceiling".* Fortunately, you notice, someone has sketched this construction on the office copy of the overall section which was faxed to site (see *Sk616*).

Question 1: As you can see from the section (*Sk616*), there is to be an overhang of approximately 750mm from the end of truss, and the fascia is to be a curved, *'bull-nose'* profile. Sketch at 1:5 scale (on **A3** paper) this lower canopy end / gutter detail (indicated on *Sk616*), paying particular attention to the following aspects:

- How the overhang is achieved structurally.
- How the roof build-up layers are terminated.
- Interaction between gutter and roof elements (structure, standing seam roofing etc)
- How the gutter, bull-nose fascia etc are supported.
- The aesthetics of the overall detail.

Tip: Enlarge the detail from the 1:100 sketch on the photocopier, and use this as a basis for the detail. Portray the elements as sloping rather than curved, for ease of drawing. Do not forget that the sketch of the roof build up (on Sk 616) is shown in a 90 degree different plane to your detail.

Question 2: In a rare coffee break, flipping through the drawings, it suddenly dawns on you that insulation in an external canopy should not be necessary. Why, then, is it there (in 10 words or less)?

Submission & Format
This exercise needs to be handed into (place?) by (date, time?)

The work should be to the following format:

- A3 ***photocopy*** – not original.
- Construction sketching technique – not computer or drawing-board drafted drawings.
- All details ***fully dimensioned***.
- The ***full names of all group members*** clearly shown on *all* sheets.

FIGURE 1 Example of a weekly exercise

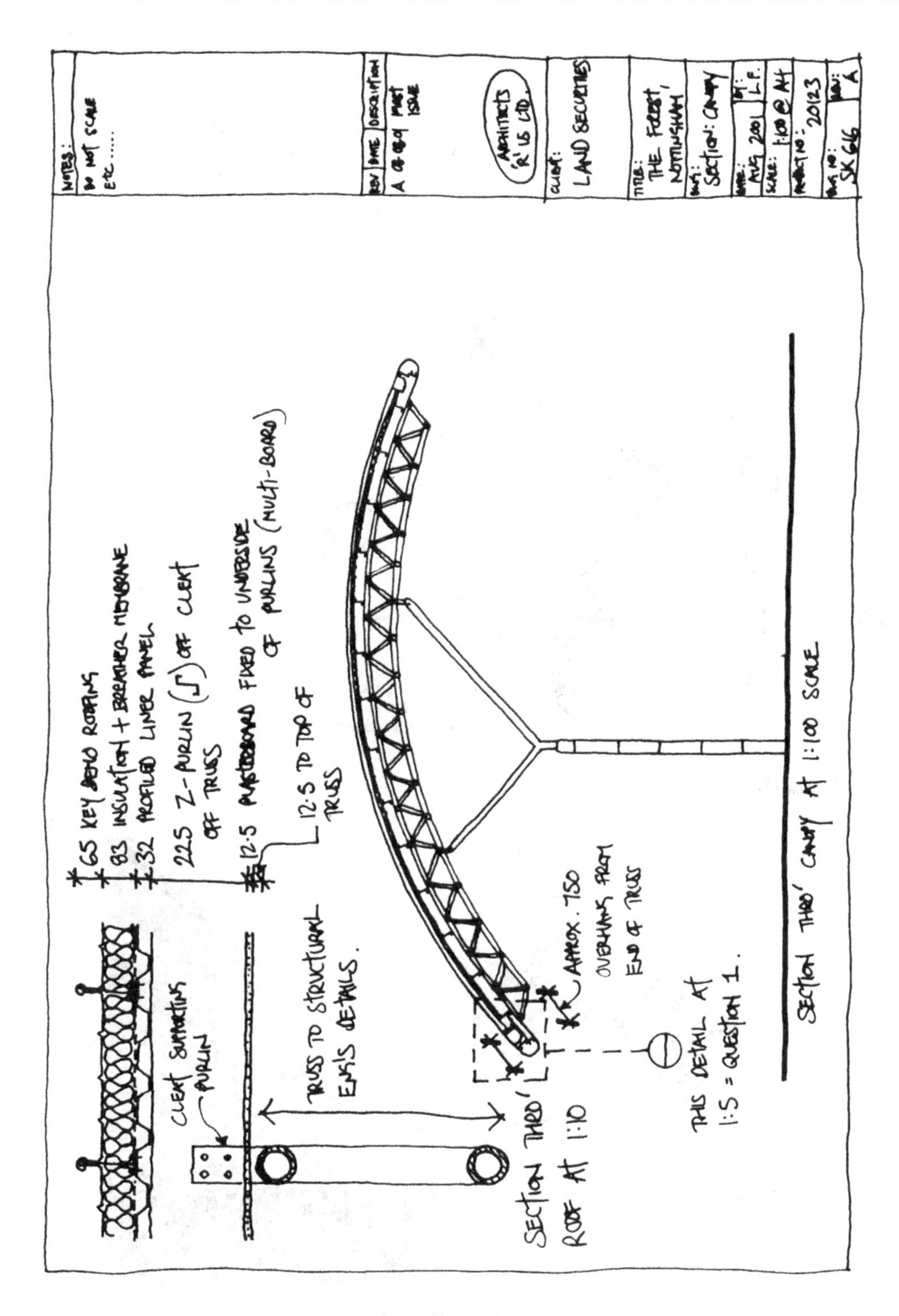

FIGURE 2 Example of supporting information for weekly exercise

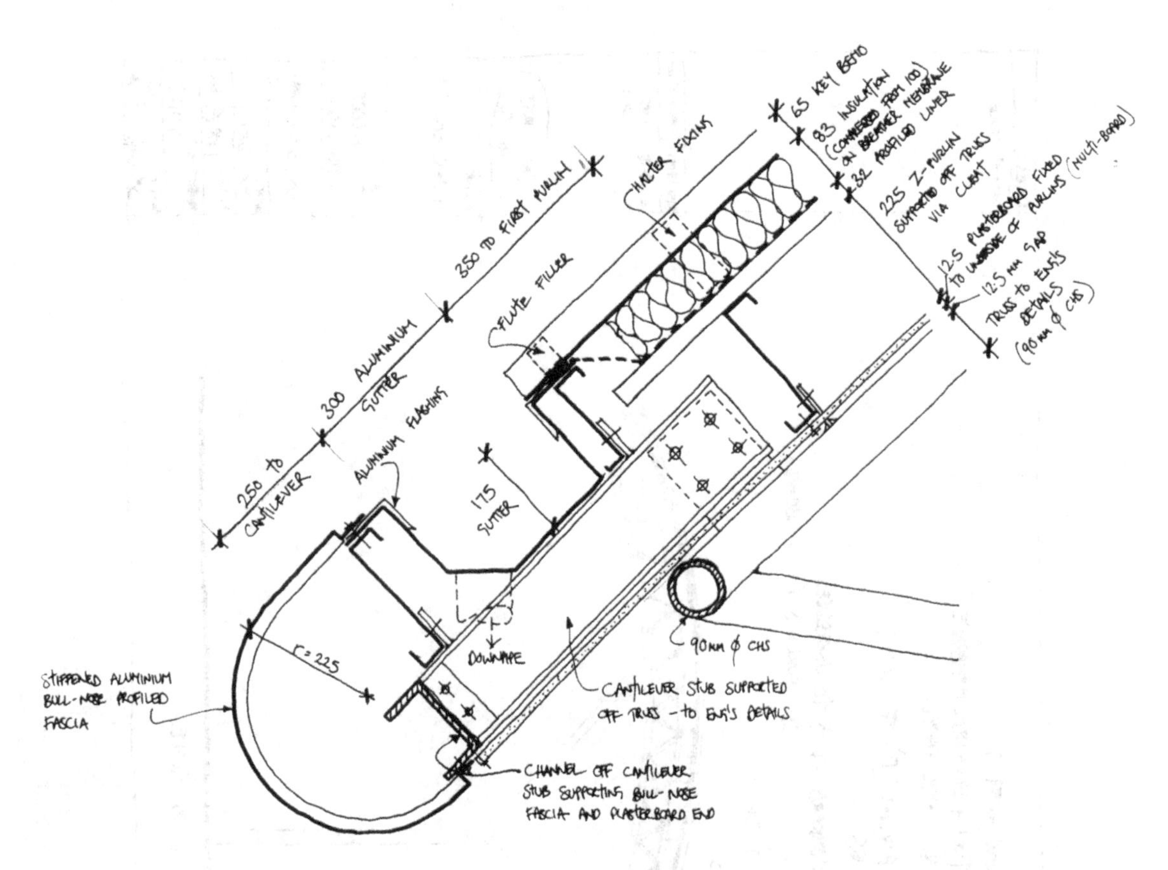

FIGURE 3 Example of response to exercise

FIGURE 4 Example of on-site construction images reinforcing exercise learning

FIGURE 5 Example of on-site construction images reinforcing exercise learning

to reinforce the link back to the continuation of the design intent from concept through to completion.

COURSE ASSESSMENT

The lecture course is assessed by an end-of-session paper introducing a building design that requires rationalizing according to its construction materials and detailing in part. Most of the details are developments of the exercises already undertaken during the lecture course, thus allowing the lessons learnt during the course to be applied directly. A large part of the paper requires the student to compile a production information 'storyboard', considering the detailed information that is needed to describe the construction of the building and how this translates into a drawing package. The finished paper is not only a collection of drawings for assessment, but also serves as a useful addition to the student portfolio for job interviews.

The actual marking of the paper adopts the same objective allocation of marks to fundamental decisions taken in the drawings as carried out in the weekly exercises. The students are also required to explain in words (briefly) why they have assembled the details as they have, to communicate the level of their understanding.

Of course, the true value of this approach to teaching construction can only really be gauged in the way that the students apply the knowledge to their own designs in studio. In third-year design studio at Nottingham we are, again, experimenting here by nestling the technology projects in the middle of design projects, rather than as an addition at the end. In requiring the student to produce structural, environmental and constructional studies during the design project, it forces the issue of technological consideration; the students cannot ignore it. The intention here is that these technological studies will help inform the design process in the latter part of the project and result in a more well-rounded design.

This twin approach in lectures and studio to technology over recent years is producing some excellent results in students' designs. It seems that students are feeling more confident about approaching construction from first principles relevant to their design, rather than irrelevant borrowing from published architects/buildings. Figures 6–8 show an example of this. This student's design involved the quite complex 'puncturing' of a massive wall element with lightweight accommodation boxes. The student worked the idea through the detailed design, developing a novel structural strategy for the wall and applying some of the lessons learnt in detailing from the lecture course directly to the design.

REFLECTION

RESULTS

The results from the student exercises in the first instance often demonstrate little understanding of the issues. Some comprehend the complexities of the system they are dealing with, but few can extrapolate

FIGURE 6 **Example of student design studio work; construction being tackled from first principles, relevant to the individual design**

this into a solution where all aspects of the detail are considered in unison. However, this does not really matter. It is in the understanding of the solution during the feedback session that the true value of the methodology lies. By attempting the exercise, the students have at least appreciated the problem and issues contained. Subsequently, in seeing the author's detail, the students understand the decisions leading to the solution. The detail has risen from an abstract plane; it has become relevant to them. In understanding what the different elements of the drawings represent and how they came to be in the positions they are, they understand the design at this most intricate scale; this particular construction has, hopefully, become demystified.

> *It is not sufficient simply to have an experience in order to learn. Without reflecting upon this experience it may quickly be forgotten or its learning potentially lost. It is from the feelings and thoughts emerging from this reflection that generalizations or concepts can be generated... The learner must make the link between theory and action by planning for that action, carrying it out, and reflecting upon it, relating what happens back to the theory. (Gibbs, 1988)*

STUDENT FEEDBACK

Feedback from the students on this particular approach to teaching construction has been extremely positive. Out of student evaluation of module (SEM) questionnaires in previous academic sessions (Figure 9), virtually all students rated every aspect of the module from five 'value' categories as either 'strongly agree' or 'agree'. In addition, a large proportion of students commented that this approach to construction teaching should be adopted in the earlier years of the course.

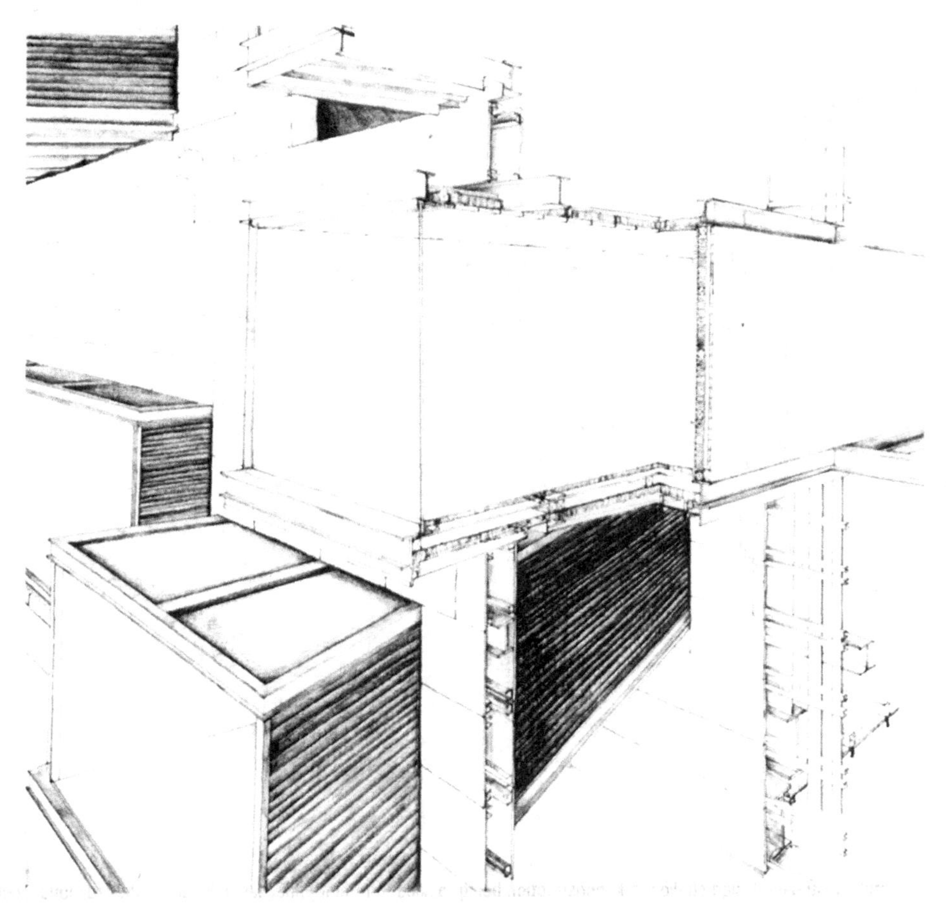

FIGURE 7 Example of student design studio work; construction being tackled from first principles, relevant to the individual design

Additional comments from the students include:

Finally, construction is taught pragmatically, realistically and vocationally in a professional manner.

A new and very much improved approach to what has previously been a rather detached subject.

The short exercises have proved to be a successful way to teach construction, much more useful than just going through things in lectures.

Great to see construction taught in such an inspirational way.

It was the first time I was actually interested in a construction lecture.

RESOURCING IMPLICATIONS

While this approach to construction teaching seems to have been successful, it is worth noting that it is resource-intensive in terms of staff input. The establishment of a course such as this requires a significant amount of time to create the weekly exercises, as well as putting the material for each weekly lecture together. Also significant in terms of resourcing, the approach can only be adopted by someone with prior practice experience to draw on (see 'Conclusion: The Way Forward?').

The most time-intensive aspect of this teaching approach, however, is the marking of the weekly exercises

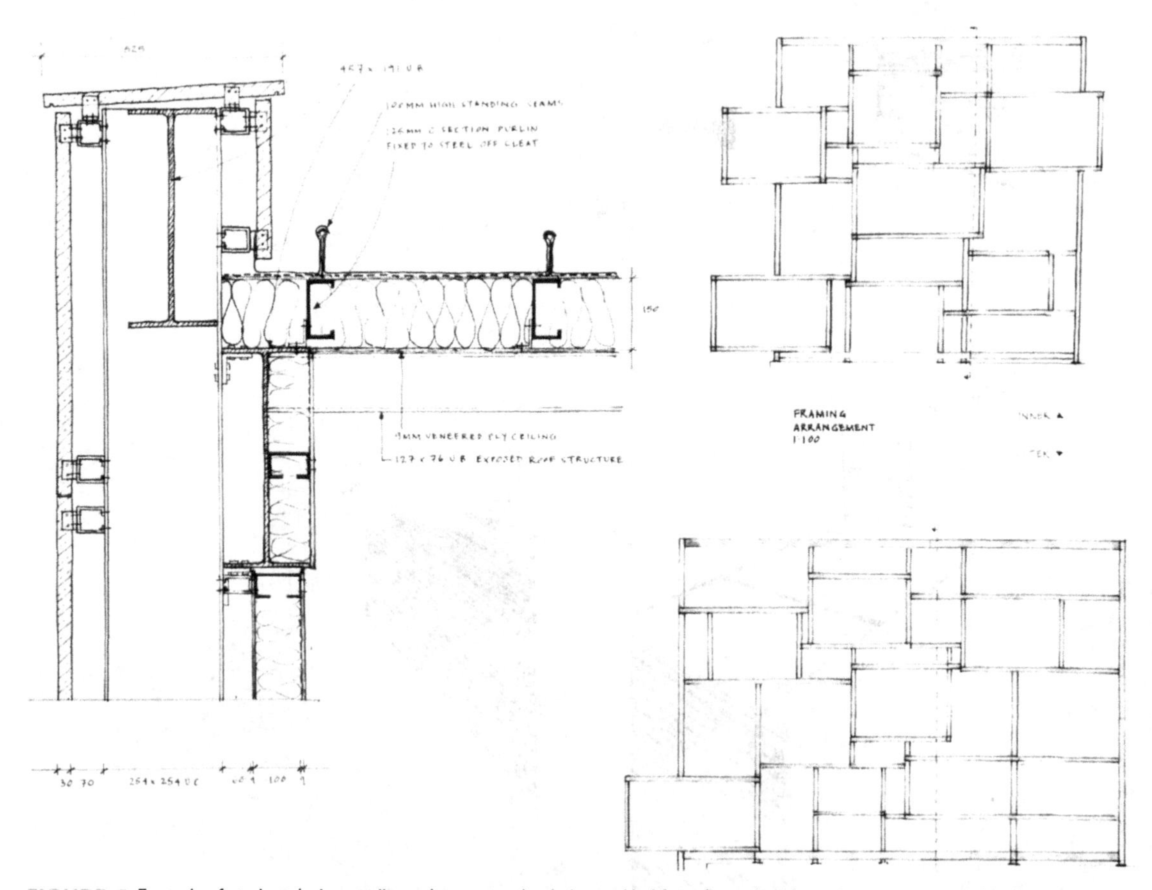

FIGURE 8 **Example of student design studio work; construction being tackled from first principles, relevant to the individual design**

(although this burden has been reduced somewhat this year by organizing the students to undertake the exercises in groups of four, which has the additional merit of increasing learning through collaboration).

Marking of the exercises is essential. As academics, one of the major pressures facing us is the fact that the majority of students only take an educational activity seriously if it is assessed. These students are extrinsically motivated by the level of importance perceived in a task, which is reflected in the amount the task grade contributes to their overall degree classification. Thus if an activity/exercise is not assessed, often students will give it less, if any, significance since it fails to bring a reward. The learning of these students occurs in a 'strategic' or 'achieving' manner (Entwistle and Ramsden, 1983).

> *Students adapt to the requirements they perceive teachers expect of them. They usually try to please their lecturers. They do what they think will bring rewards in the systems they work in. All learners, in all educational systems and at all levels, tend to act in the same way. (Ramsden, 1992)*

This attitude towards marking (or the perceived value of parts of an educational course in terms of its contribution to a qualification classification rather than actual learning) influences what Ramsden (1992) has termed 'surface' and 'deep' approaches to learning. A 'surface' approach to construction learning involves the copying of previously published details without regard for appropriateness or understanding of content, or the memorizing of information from lectures for

SET Questionnaire — Page 1 of 1

Student Evaluation of Teaching

Mr Antony Wood
Construction Design 3
K1CCD3
Lecture
University Park

Built Environment Session 02/03

This questionnaire is part of a continuing effort by the University to improve teaching and promote learning. Your responses are anonymous and will feed directly into plans for teaching enhancement and staff development.
Answers to Sections A and B are analysed by the University. Please answer all the questions in these sections either by ticking the category which best reflects your view or by ticking Not Applicable (N/A).
Section C is your opportunity to provide feedback in your own words and is returned to your teacher.

A: Fixed Questions Set by the University

		Yes	No
1	Is this module compulsory for you?	✓ 1	2
2	I have attended at least 80% of the sessions timetabled for me with this teacher	✓ 1	2

		N/A	Strongly Agree	Agree	Neutral	Disagree	Strongly Disagree
3	The teacher was an able communicator	1	✓ 2	3	4	5	6
4	The teacher retained my interest	1	✓ 2	3	4	5	6
5	The teacher was approachable	1	✓ 2	3	4	5	6
6	Sessions were paced appropriately	1	✓ 2	3	4	5	6
7	Overall, this teacher assisted my learning	1	✓ 2	3	4	5	6

B: Questions Selected by the School

		N/A	Strongly Agree	Agree	Neutral	Disagree	Strongly Disagree
GEN011	The teacher makes good use of examples and illustrations	1	✓ 2	3	4	5	6
GEN014	The teacher emphasizes key points	1	✓ 2	3	4	5	6
GEN021	The teacher sets high standards	1	✓ 2	3	4	5	6
GEN039	The teacher seemed to know the subject well	1	✓ 2	3	4	5	6
GEN042	I have been encouraged to take responsibility for my own learning	1	✓ 2	3	4	5	6
GEN067	The teacher points out links to previous topics we have discussed	1	✓ 2	3	4	5	6
GEN090	Although difficult, I understood this subject in the end	1	✓ 2	3	4	5	6
GEN104	The overall subject matter was developed logically	1	✓ 2	3	4	5	6
MAT005	The handouts helped me to understand the material	1	✓ 2	3	4	5	6
MAT007	The teacher made good use of audiovisual materials	1	✓ 2	3	4	5	6
MAT034	I used a textbook(s) in addition to lecture notes	1	2	✓ 3	4	5	6
VOC002	I have learned the relevance of this subject to my future profession	1	✓ 2	3	4	5	6
GEN013	The teacher structures the material well	1	✓ 2	3	4	5	6
GEN019	The teacher points out links to other subjects	1	✓ 2	3	4	5	6
GEN036	The teacher gave explanations which were clear	1	✓ 2	3	4	5	6
GEN049	The teacher is available for consultation	1	✓ 2	3	4	5	6
GEN051	My ability to work independently has improved	1	✓ 2	3	4	5	6
GEN086	In this module I was encouraged to think	1	✓ 2	3	4	5	6
GEN043	The teacher makes good use of class time	1	✓ 2	3	4	5	6

FIGURE 9 **Example of student feedback on teaching approach**

regurgitation, again without much understanding. The author believes that the approach outlined in this paper encourages a 'deep' approach to learning, which is borne out in both the quality of the work produced by the students and their employability as architectural graduates.

THE CASE FOR PRACTICE: COMPLIANCE OR CONTESTATION?

Architectural academics are not the only ones concerned with the lack of ability of students to integrate technology into design. If anything, the profession and the building industry are even more vociferous in these concerns (Rashleigh, 2001) and are increasingly calling into question the preparation of graduates for the commercial and practical realities of working practice.

> *Practitioners are expressing disappointment, if not outright disgust, about how well students are trained for the profession... While no one expects graduates to detail expertly or manage a project on their first day on the job, practitioners observe that many graduates have little awareness or appreciation of these skills, among scores of others important for the creation of architecture. (Crosbie, 1995)*

This raises the question of whether we should be preparing students for the realities of working practice at all? An important debate across the higher education spectrum is what Rowland (2001) has termed the question of 'compliance or contestation' in tertiary education – should we be training graduates in accordance with the skills that the 'real' world of commerce or industry requires of them or should we be fostering in them a more personal line of development, giving them the opportunity to explore and nurture some of the talents that are perhaps lying latent within, which the pressures of the real world, once entered, would rarely allow? Are we here purely to prepare graduates according to the wishes of the commercial world, or should we be investing that energy in helping the students to challenge that view of the world, and challenge themselves? Certainly a curriculum obsessed with attaining skills does not seem conducive with one that fosters freedom for the student to explore, expand and experiment which are all vital components of an architectural education.

The pressure on students to perform in terms of grades is now such that most seem to rely only on exploiting their strengths, rather than developing their weaknesses. The concept of experimentation is being rejected by many students in favour of 'safe' options, in the hope of obtaining higher grades. University seems now to be more about obtaining a qualification than an education. When the realization that what the student develops into during the education years – what he/she takes into the heart and head – is far more important than any grade on paper, an opportunity has often sadly been lost.

So should we be training students primarily for the job that many of them (but by no means all) will eventually be doing, or should we be helping them to develop personally in directions that may not even be architectural? In academia, the response to the accusation of irrelevance to the profession usually manifests itself in the retort that university is not there to replicate practice; that the architectural office is the proper place for training and that university is concerned primarily with helping students attain a higher level of philosophical understanding – certainly in terms of design appreciation – perhaps before the ravages of a career in architecture whittles away the ideal. Many schools of architecture thus place little emphasis on the teaching of construction technology, instead focusing on design concepts, aesthetics and the theory of architecture, leaving the practitioner to teach the necessary technical skills required to design buildings in the real world. This debate about the direction of architectural education has manifested itself in the consideration of whether architecture is a 'discipline' or a 'profession', an 'intellectual pursuit' or a 'vocation' (Crosbie, 1995).

I do not see, however, why these two visions of architecture (or, more specifically, the graduate) have to be necessarily at odds; why one has to be to the detriment of the other. What seems to be missing from the argument is an element of perspective. Would, for example, teaching construction to students following the methodology outlined above – with the dual purpose of helping them to understand the essential link of construction to the design process as well as helping them to prepare for practice – really dampen down their ambition, or somehow corrupt their ability to appreciate the beauty of the philosophy of design? For those that believe that the approach to construction teaching outlined above is indicative of a retreat to the philosophical dark ages, it should perhaps be pointed out that this particular construction course constitutes approximately 8% of the third year; more than 50% of the overall course is still focused on design studio.

CONCLUSIONS: THE WAY FORWARD?

The pressures on the content of the architectural curriculum (keeping pace with the practice of architecture where responsibilities seem to be increasing exponentially) are now so great that we cannot afford to lose the opportunities proffered to us by each element of it. We need to be excellent at each element that we teach, be it philosophy or construction, history or building services. Each course needs to be crafted to maximize its contribution to the student's learning.

> *Curricula should be designed in order to present students with experience of architecture as an integrated form of knowledge, in contrast to the present practice of planning additive collections of courses which provide different and uncoordinated fragments of knowledge. (Abel, 1981)*

In construction, certainly if the ranks of architectural professions are to be believed, we are not achieving that. So, why not? If the teaching of construction is elementary and methodologies such as those outlined above have been known for years, why, as schools of architecture, are we failing to implement them? This paper does not pretend to be revolutionary as an approach to teaching construction in schools of architecture. Presenting specific drawn details and reinforcing this with images of on-site operations, or providing physical material samples for students to experience is not a new concept, nor is the idea of encouraging learning through problem-based exercises. But, somehow, these – and other – methods of relevant construction teaching must be getting lost, since architectural students generally seem unable to adequately incorporate technology in their designs.

I believe that a significant factor in this is the lack of actual building experience among architectural academics. A prerequisite for the teaching of construction is the prior experience of building buildings. This brings us back to the problem that opened this paper; tutors can explain the theoretical links between construction and design, but very few can actually demonstrate why details are put together in the way they are.

> *Professional education consists of passing on a body of knowledge and skills. Students ... need to be inducted to the discipline in a special way – by direct learning over the drawingboard from people who have themselves designed and put up buildings, and can pass on their skills. (Davey, 1989)*

The gap between academia and practice is widening. Each week, a legion of practitioners enters our schools of architecture to assist in part-time teaching but, from academia, this interchange is not reciprocated. Due to pressures imposed on us by structures of higher education such as the Research Assessment Exercise (RAE), the number of academics involved in the building of buildings is very small. And thus, construction and other subject areas are taught by academics with little relevant experience, or perhaps experience garnered so long ago that it is all but irrelevant.

> *One of the areas for concern must be the continued drift away from institutions' teaching of technology. The colleges have got to shift the focus of teaching by having tutors with experience and interest [in these areas]. There are some heads of school who do not have a sufficient understanding of practice. (RIBA Vice-President for Education, Paul Hyett, in Booth, 2000)*

This lies at the crux of the problem and it is here that we need to address it, coupled with creating courses that maximize the learning potential. It will not be easy. Pressures on both academics and practitioners are such that a retreat into their own world, to cope with those pressures, is increasingly inevitable. But we need to foster greater links between the two strands that constitute architecture in this country – a greater dialogue, more cross-over, increased collaboration.

In a similar way that all practitioners are required to undertake continuing professional development (CPD) during their careers, perhaps all academics – and especially those involved in the teaching of construction and related areas – should be required to spend periods in practice, on site, etc.; a kind of reverse CPD.

An idea similar to this has been implemented by Kingston's School of Surveying and Architecture and

Landscape. The programme 'Learning to Work: Working to Learn' encourages teaching staff to take a two-week sabbatical in an architects' practice where they will shadow a particular member of staff, sit in on relevant meetings and visit sites. The programme aims to act as both a refresher course for academics who may have grown rusty about current practice issues and as an ideas source for promoting teaching within the school with more relevance to practice. The scheme has proved popular, with 16 academic staff involved and five architectural practices providing the necessary experience (Williams, 2000). The idea of the Architecture Professional and Educational Centre (APEC) has also been suggested, providing a place for communication and collaboration between academia, practice and the community (Musgrove, 1978).

There will be many ways of achieving closer collaboration, but the dialogue has to improve. We need a greater understanding of what the other does and a retreat from the insecurities that leave no option but to 'fight the corner'. Both academia and practice need a greater common ground for the benefit of the graduates who are the future of architecture in the UK.

> *Ultimately, more radical educational and administrative solutions will be required to bring architecture schools in line with cutting-edge practice. Most important, cooperative links with industry and practice need to be recognized, not as a restriction on creative freedom ... but as a means of helping students and staff alike to master the skills which make such freedom possible. (Abel, 1998)*

AUTHOR CONTACT DETAILS

Antony Wood: School of the Built Environment, University of Nottingham, University Park, Nottingham, NG7 2RD, UK.
Tel: +44 (0) 115 951 3111, fax: +44 (0) 115 951 3159,
e-mail: antony.wood@nottingham.ac.uk

REFERENCES

Abel, C., 1981, 'Function of tacit knowing in learning design', first published in *Design Studies*, 2(4), 209–214, October 1981. Now in Abel, C., 2000, *Architecture & Identity: Responses to Cultural and Technological Change*, 2nd edn, Oxford, Architectural Press, 106–113.

Abel, C., 1998, 'Networking the studio: architectural education and the virtual practice', first published in *Environments by Design*, 2(1), 101–116, Winter 1997/98. Now in Abel, C., 2000, *Architecture & Identity: Responses to Cultural and Technological Change*, 2nd edn, Oxford, Architectural Press, 68–77.

Booth, R., 2000, 'New graduates are not up to scratch', in *Architects' Journal*, 212(17), 12.

Carpenter, W.J., 1997, *Learning by Building: Design and Construction in Architectural Education*, New York, Van Nostrand Reinhold.

Chadwick, M., 2004, 'Back to school: architectural education – the information and the argument', in *Architectural Design*, 74(5), 4–102.

Cook, P., 2004, in Chadwick, M., 2004, 'Back to school: architectural education – the information and the argument', in *Architectural Design*, 74(5), 6–12.

Crosbie, M.J., 1995, 'The schools: how they're failing the profession (and what we can do about it)', in *Progressive Architecture*, 76(9), 47–51, 94, 96.

Davey, P., 1989, Architectural Education special issue, *Architectural Review*, 185(1109).

Dawson, S., 2000, *Architect's Working Details (Volumes 1–9)*, London, *Architect's Journal* in conjunction with Emap Construction.

Entwistle, N. and Ramsden, P., 1983, *Understanding Student Learning*, Kent, Croom Helm Ltd.

Gibbs, G., 1988, *Learning by Doing: A Guide to Teaching and Learning Methods.* London, Further Education Unit.

Hirst, P.H., 1975, 'Liberal education and the nature of knowledge', in R.S. Peters (ed.), *The Philosophy of Education*, Oxford, Oxford University Press, 87–111.

Musgrove, J., 1978, 'Schools of architecture and the profession', in *Architects' Journal*, 167(24), 1153–1158.

Ramsden, P., 1992, *Learning to Teach in Higher Education*, London, Routledge.

Rashleigh, B., 2001, 'Schools go back to crisis', in *Building Design*, issue 1500, 1–2.

Rowland, S., 2001, 'Is the university a place of learning? Compliance and contestation in higher education', inaugural lecture, University College London.

Williams, A., 2000, 'Teaching old dogs new tricks', in *Architects' Journal*, 212(18), 42–43.

ARTICLE

Teaching and Learning in Collaborative Group Design Projects

Richard Tucker and John Rollo

Abstract

The 200 years of apprentice/master tradition that underpins the atelier studio system is still at the core of much present-day architectural design education. Yet this tradition poses uncertainties for a large number of lecturers faced with changes in the funding of tertiary education. With reductions in one-to-one staff/student contact time, many educators are finding it increasingly difficult to maintain an atelier teaching model. If these deficiencies remain unchecked and design-based schools are unable to implement strategies to reduce the resource intensity of one-to-one studio teaching programmes, then, for many higher-education providers, current architectural education may be based on an untenable course structure. Rather than spreading their time thinly over a large number of individual projects, an increasing number of lecturers are setting group projects. This allows them to coordinate longer and more in-depth review sessions on a smaller number of assignment submissions. However, while the group model may reflect the realities of the design process in professional practice, the approach is not without shortcomings as a teaching and learning archetype for the assessment of individual student skill competencies. Hence, what is clear is the need for a readily adoptable andragogy for the teaching and assessment of group design projects. The following is a position paper that describes – with a focus on effective group structures and assemblage and fair assessment models – the background, methodology and early results of a Strategic Teaching and Learning Grant currently running at the School of Architecture and Building at Deakin University in Australia.

■ *Keywords* – Collaborative learning; group assessment; design andragogy

ESTABLISHING BEST-PRACTICE PRINCIPLES FOR THE TEACHING AND ASSESSMENT OF GROUP DESIGN PROJECTS

BACKGROUND – DRAWING ON THE BATH AND CUDE MODELS

Architects need collaborative skills to negotiate infinite design options within a building design process that can include more than 50 types of participants and consultants (Cuff, 1991). Yet, while a significant body of research exists relating to the teaching of problem-based group work (Dillenbourg, 1999; Sanz-Menendez *et al*, 2000; Grigg *et al*, 2003), the focus of this research has rarely been the student design studio. Only the Clients and Users in Design Education (CUDE) project at Sheffield University School of Architecture in the UK has looked at the issue of teaching teamworking skills to students of architectural design. The findings of the CUDE research are, however, untested elsewhere as a measure of student performance in team design in comparison with performance in individually assessed projects. Investigations at Deakin are attempting to redress this shortcoming by observing, recording and analysing student performance and feedback in group and in individual design programmes.

Group project working is, of course, not new to architectural education and has been a core aspect of many multidiscipline or joint schools. One such programme that both of the authors – Tucker and Rollo – worked with during the mid 1990s that highlighted many of the benefits of collaborative learning (an instructional approach in which students work together in small groups towards a common goal (Dillenbourg, 1999)) is the model established by Professor Ted Happold at the School of Architecture and Civil Engineering at Bath University in the UK (School of Architecture and Civil Engineering, 2004). While the school operates separate course structures for the two disciplines addressing the requirements of the different accrediting professional bodies, they have components within each of the four undergraduate years where both discipline cohorts combine in a studio environment to engage in joint project work. One of the programme units manipulates groups to develop collaborative student teams consisting of at least one architecture student and at least one engineering student. These students are subsequently tutored at the same time as a single collaborative team by a pair of supervisors comprised of one architect and one engineer. Teams work together on the brief from concept, through schematic design, to design development, and submit a single 'seamless' joint submission. Although most schools of architecture are single discipline focused, the authors strongly believe that, if managed appropriately, many of the benefits gained by having students working jointly – such as greater breadth and depth of knowledge acquisition, more thorough design exploration and resolution, and more accomplished presentation – can all be translated into a single discipline curriculum.

Tucker and Rollo now operate as coordinating lecturers of the third- and fourth-year studios, respectively, at Deakin University and have introduced collaborative group projects to their cohorts. However, although the collaborative model works well in some instances, both authors are equally aware of situations where it has failed. Given rapid changes that have come about in the funding of tertiary education (which places significant pressure on one-to-one staff/student contact time), an increasing number of schools and educators are being compelled to operate group projects. If, therefore, collaborative learning is to become a standard studio teaching model, it is imperative to establish best-practice teaching standards in order to optimize group dynamics, the breadth of shared experience, the transfer of knowledge and skill base, and fair assessment criteria. The following is a position paper that describes the background and methodology of a funded research programme currently running at Deakin University to address these issues.

STRATEGIC TEACHING AND LEARNING GRANT

Members of Deakin School of Architecture and Building were recently named as recipients of a Strategic Teaching and Learning Grant (STALG) aimed at 'establishing best-practice principles for the teaching of group design projects'. The project builds on collaborative research between the school and a Deakin University teaching and learning support service (Deakin Learning Services) that, in 2004, identified the need for additional resources to assist in group teaching. The STALG-funded project is evaluating two design programmes at Deakin – the third-year Atelier Geelong studio and the fourth-year Urbanheart Design Research Forum – and it is hoped that the findings will inform an andragogical framework that at present does not exist for design teaching. It is envisaged that the results of the investigation may also inform other project-based teaching disciplines experiencing a similar need for new knowledge and skill-based delivery due to increasing staff–student ratios.

The outcome of the research primarily focuses on results that can be measured in student achievement as reflected in four indicators including grades and graduate outcomes, satisfaction as reflected in student evaluations, knowledge and skills gained through the measure of student design projects, and the effect of the make-up and selection of groups on the decision-making processes as assessed by studio observation and informed by the results of Myers Briggs-based personality type testing that has a long history in educational literature as an effective way of engineering productive group environments (Durling *et al*, 1996; Chambers *et al*, 2000; Chang and Chang, 2000; Rutherfoord, 2001; Gorla and Lam, 2004; Deibel, 2005).

An additional focus of the Deakin research addresses the often overlooked question of assessment. In the experience of design teachers at Deakin, the issue of 'fair' assessment is one of great concern to academics and students alike in team design projects – the success of which often hinges on students' perceptions of assessment reflecting their comparative performance. The project, therefore, is implementing and evaluating online team assessment methods that are being developed to allow students to assess one another's performance in a group within the secure and anonymous environment of a web portal.

The project aims ultimately to inform a change of classroom/studio practice governing the assemblage, teaching and assessment of design teams. This also includes the development of adult-learning principles focused on effective collaboration and fair assessment leading to an enhanced student group learning experience, which should subsequently enhance studio learning and professional practice.

ATELIER GEELONG

The beginnings of the STALG group learning project were prompted by a situation likely to be all too familiar to those teaching design. In 2003, the 95-strong cohort taking part in the first semester third-year design studio could expect a maximum of 10 minutes per week one-to-one teaching time. In common with many other schools across Australasia, Deakin students that were set individual design projects could not rely solely therefore on one-to-one contact time with tutors to advance their solutions through the iterative process of reading, questioning, testing and reformulating that characterizes the design process. There is, of course, one easy solution to this problem and that is for tutors to review fewer assignments, but in greater depth, by setting group design projects.

Two major design projects were established for the 2003 first semester studio. To use scarce teaching resources efficiently, the first was a team design project taught largely by two tutors individually overseeing groups of six to nine students, and the second was an individual project taught by four tutors seeing individual students for 20-minute one-to-one tutorials. The third-year students were asked therefore to divide themselves into groups of three for their first major design project – Atelier Geelong. As the Atelier programme is the prime focus of the 2005 STALG research project, we shall look at its evolution in some detail.

The Atelier would be designed by students to provide living accommodation and studios for Geelong graduates mastered by a tutor, with a split tenancy arrangement which allowed for a scheme that could readily be subdivided into three distinct elements. The brief concluded:

> *Of course, the design of your Atelier might counter this subdivision or even further it. This is unimportant, what is important is that at around three to four weeks into the project the design team must break their submission and presentation, and hence the focus of each individual member, into three separately appraisable elements.*

This requirement underlines the problem of many a team design project, for what is commonly desired is one design solution that reads as consistent and seamless, but one that allows for the separate appraisal of those who devised it. And of course this – the best of both worlds – is difficult to achieve and is fundamentally conflicting. The question of appraisal is one that shall be revisited throughout this paper.

THE COLLABORATIVE STRUCTURES OF ATELIER GEELONG TEAMS

Tutors can allow students to self-select teammates or can allocate them to specific groups. Allocated groups can then either be randomly assembled or engineered to create teams containing a range of experiences and abilities. Each method has its own merits. Although some students may prefer to be allocated to a group and may view this as fair, particularly if the groups are randomly selected, experience at Deakin has shown that self-selecting groups are usually more popular for they minimize personal conflict and reduce the need for tutor intervention in disputes. It was for these reasons that students were allowed to choose their own teammates for the 2003 Atelier project.

The teamworking of approximately 40% of the 2003 teams can be described with the term 'democratic collaboration'. This resulted when there was no clear leader and/or, in many instances, when students were too polite or of such similar ability that they were disinclined to criticize in depth. In such cases, all ideas were treated as equal, meaning that those developed were those elected democratically. This often implied that the ideas selected had prompted the fewest objections, which frequently resulted in a product that in advertising parlance is commonly (unkindly) known as the 'lowest common denominator'. This clearly was not a mode of collaborative working that encouraged risk for, as Schrage (1995) implies, innovation is more often than not the product of a diverse range of skills and abilities. The problem of hierarchy within a group, or rather the lack of it here, is one that this paper will return to, for it is significant.

It can be seen from the analysis of past results that, in 2003, groups of the democratic type were comprised largely of students of similar abilities. Indeed, the reason why this type of group accounted for the majority seems to be that students, when allowed to choose whom they worked with, chose like-minded peers with similar levels of skill. The democratic collaborators could therefore be split into three subgroups – the high, low and the average achievers as defined by their previous project marks. The average achievers, numbering seven, were the most widespread type of democratic group. There were two groups of high achievers, including one group comprised entirely of students who had in the previous semester achieved at least a 70% grade or higher. The high quality of work produced by these two teams was in stark contrast to the work produced by the five low-achieving teams, comprised largely of students who had merely achieved pass grades the previous semester. Indeed, four of these low-achieving groups failed the project and the disparity between their work and that of the higher achievers became the source of discontent. When the low-achieving groups became acutely aware of how far they were behind the high achievers at an interim review, their already lesser motivation was seemingly reinforced for some who saw the skills imbalance as an insurmountable disadvantage. Their initial frustration at their own perceived lack of ability became externalized to discontent with the course and programme structure, feeding, therefore, into a notion described in the field of social psychology as the 'self-fulfilling prophecy in the classroom' (Smith and Mackie, 2000; Deibel, 2005). It was decided in future to avoid putting low-achieving students in what they themselves clearly identified as a situation that amplified their weaknesses rather than addressing them.

It might be appropriate to describe the groups driven by one or two high achievers, which numbered eight – the least common of the three primary collaborative modes – as 'oligarchic collaborators'. Not only did these groups often produce the most accomplished and innovative designs, but they usually resulted in a positive learning environment for everyone. This included low achievers, who in these groups were often encouraged to develop previously unchallenged abilities. Here, hierarchy within the group was certainly of benefit.

If 40% of the teams could be described as democratic and 25% as oligarchic then, in turn, to describe the organization of approximately another 35% of the teams we might use another term with Platonic origins, namely 'timarchic collaboration'. For, in common with Plato's description (translated by Lee, 1955) of a society divided by internal strife and characterized by conflict and selfish ambition, this last type of group was born out of dissent. Often the result was piecemeal design with little cohesion – a kit of parts with little to unite it. Most failures of teams to bond, due to either clashing egos or other failures to communicate, led to this common solution; namely, the piecemeal design of disparate parts defined merely by an allocated footprint. The problem was exacerbated by the demand in the 2003 brief that each student must develop a 'separately appraisable element'. It was subsequently decided that for 2004 the requirement for separately appraisable work would only be introduced towards the end of the project – when presentation became the focus, because presentation by its very nature demands the allotment of tasks.

The allocation of students to larger teams of six or seven in 2004, engineered to contain a range of different experiences and abilities, at once evened out the significant disparity in the quality of the teams' work that in 2003 had so undermined the self-confidence of low-achieving students. Indeed, in 2004, no teams failed the group project Atelier Geelong. Yet this significant victory was achieved at great cost, for the distribution of the three modes of group collaboration saw, in 2004, a large increase in the proportion of timarchic teams. Grouping strangers rather than friends led to much more conflict. The experience nurtured a perception that the engineering of groups benefited not the students but rather the tutors – who, students felt, merely wished to avoid dealing with 'remedial' groups.

This overshadowed the feeling of a small but significant number of students that the diversity and sometimes the inherent hierarchy of a mixed ability group enhanced collaboration. As one student put it in a trenchant summing up of working in allocated groups (Anderson, 2004: 23–24):

> *Has working in an allocated group been a positive or negative experience? Both, because I may not have enjoyed any of the experience, finding it stressful and upsetting due to group dynamics and personality clashes but it was a positive learning experience about how to work with people that you don't know or get along with or don't respect you. I also learnt a lot about my own flaws and ideas that I can bend and adapt to work better with people in the future.*

The mixed feelings for the 2004 Atelier project can be summed up in an unpublished student support body (Deakin Student Life) survey in which 77% of the cohort stated that the group project had offered positive experiences of group working, while 54% of the cohort identified 'choosing your own group members' as the change that would make the experience more positive. Moreover, 52% of the cohort identified reducing the size of the group as the next most popular change, for the problems of collaborative designing, at least in the school studio, were seemingly exacerbated in larger groups.

In response to student feedback, the 2005 Atelier project will ask the cohort to self-select themselves into groups of five. However, diversity in these groups shall be encouraged by adding controls to this self-selection process, for students will be asked to form groups within diverse pools that may not necessarily contain their friends, but will instead be assembled according to diverse personality natures identified by a Myers Briggs-based test. This type of test has already been popular among students within the faculty as a way of identifying how different team-role preferences can benefit collaborative working. It remains to be seen whether the best of both worlds of self-selection and group diversity is indeed achievable.

The responses of two students in 2004 sum up a problem voiced regularly relating to the lack of hierarchy within groups (Anderson, 2004: 25, 28).

> *If it's not on the students' part to choose their team members, there's a whole lot of problems. In a working atmosphere, people have different posts so it doesn't cause problems.*
>
> *I find that I work a lot better on my own. My own skills improve when it's my own project. Working in groups in a [professional] work environment is a lot better. It has structure and a definite leader (you also have designated working time together i.e. 9–5pm).*

The lack of a natural hierarchy within the groups was found to be important for two reasons. The first relates to a perception touched on by both of these students and echoed by many of their peers, for although 68% of students saw the purpose of the Atelier project as preparation for employment, many saw it as an unrealistic construct of professional practice, because practitioners, unlike themselves, have clearly differentiated and ordered roles. This perception not only undermined the credibility of the group project as a mode of teaching the collaborative skills needed to negotiate the building design process, it also undermined the collaboration of the groups themselves. For, without clear leadership, it is often impossible for a team to make decisions, or rather the decisions that are made are merely those that offend the least. Most students recognized that these 'lowest common denominator' decisions were not the best. Although the allocation of students to diverse groups sometimes led to the identification of a dominant leader or leaders, and these oligarchic groups in both the 2003 and 2004 semesters produced the best work, this was not the case for the majority of groups. Thus, in 2005, all teams will be asked to elect a leader or spokesperson, but the focus on this leader will be made less sharp by providing a leader at the next level of hierarchy in the design tutors themselves. To avoid leading the design process, tutors will merely adopt a role at the head of client groups in their meetings with their design groups, to identify the ideas that might be developed rather than suggesting them. This directs us to a final conclusion about role definition and one concerning the nature of the role of the teaching staff. When asked by Deakin Learning Services what would make group work a more positive experience, 35% of students stated in 2004 that greater staff support in the process of collaboration itself was necessary. As tutorial times in 2005 will be allocated to coincide with design/client group meetings, it is hoped that tutors will be able not only to act as mediators in design decisions but, moreover, play an overseeing role in conflict resolution – a role for which many students seek support.

ASSESSMENT

It became clear early in the 2003 Atelier project that the more conscientious students were aggrieved by what they saw as an inequality in their workload. As one student complained, 'it is easy to free-ride in a group and, unfairly, it is us, the hard workers, that have to carry the lazy ones'. It became clear that a mechanism would have to be built into the appraisal of the projects that rewarded those working hard while penalizing those that were not (McGourty *et al*, 2000). It was therefore made clear to students that after the final review they would be asked to assess each others' contribution to the team. This system appeased those who felt aggrieved, but suffered from one major problem, namely that the adjustment in grades from only one peer/self-assessment process could overcompensate. If a number of students were feeling particularly vindictive, their exaggerated misallocation of marks could unfairly penalize teammates. A solution to this problem that was pursued in 2004 was continuous peer/self-assessment on a weekly basis, to distribute evenly anomalous or unfair scoring across the semester.

Although continuous peer/self-assessment in 2004 solved the unrealistic allocation of marks, fair assessment is still a major concern and contributes to an overall negative perception of group teaching. Perhaps the greatest cause of these problems was the choice of peer/self-assessment that was given to teams in 2004. The students devised three options of mark allocation, either by round the table bargaining, by secret ballot or by simply allocating marks evenly. It was among the teams that perhaps rather naively chose the third option – and this was the majority – that most problems arose. In contrast, the teams that adopted the assessment methods that allowed for penalty and reward saw the allocation of marks as fair. Significantly, the vast majority of students in these teams described in their reflective portfolios the group project as a positive experience. Thus, in 2005, an online and compulsory peer and self-assessment procedure – the process by which groups of individuals rate their peers (Falchikov, 1986; Oakley *et al*, 2004) – is being developed that allows students to assess each others' contribution on a weekly basis within the secure and anonymous environment of the school intranet portal. Another consideration that may be built into the structure of this research is the possibility of reinforcing the peer/self-assessment model with a multiple response Likert scale evaluation. This Likert evaluation will allow for the coding of responses and the subsequent statistical analysis of possible patterns of bias in student responses (Ellis and Hafner, 2005).

STUDENT RESULTS

Two statistics stand out from the results of the 2003 cohort. The first can be seen from the distribution of results for the group project, namely that 40% of students achieved marks that were outside of one standard deviation from the mean mark of 55.3% (Figure 1).

This compares with the figure of 30% that is predicted by normal distribution (it should be noted here that marks are not adjusted at Deakin to distribute them according to the normal distribution bell curve). This high number of fails and of distinction of higher students equates to the large numbers of students collaborating as oligarchic groups in addition to those collaborating as high- and low-achiever democratic groups. It was decided to prevent this problematic grade distribution in 2004 by engineering teams to contain a range of different abilities.

A second notable statistic was even more prominent in 2003. Namely, that the average mark achieved by students for the group project was 3.5 percentage points higher than that they achieved for the subsequent individual design project. The studios were also less populated during the second half of semester when individual design projects were taught. This poses two questions that will be pursued in our further research: Does the performance of students as reflected in marks decrease when they work in greater isolation? Or, when the effects of group manipulation are removed, does the negative self-fulfilling effect on the lower-achieving students become increased to the extent that it pulls down the overall average grade of the cohort?

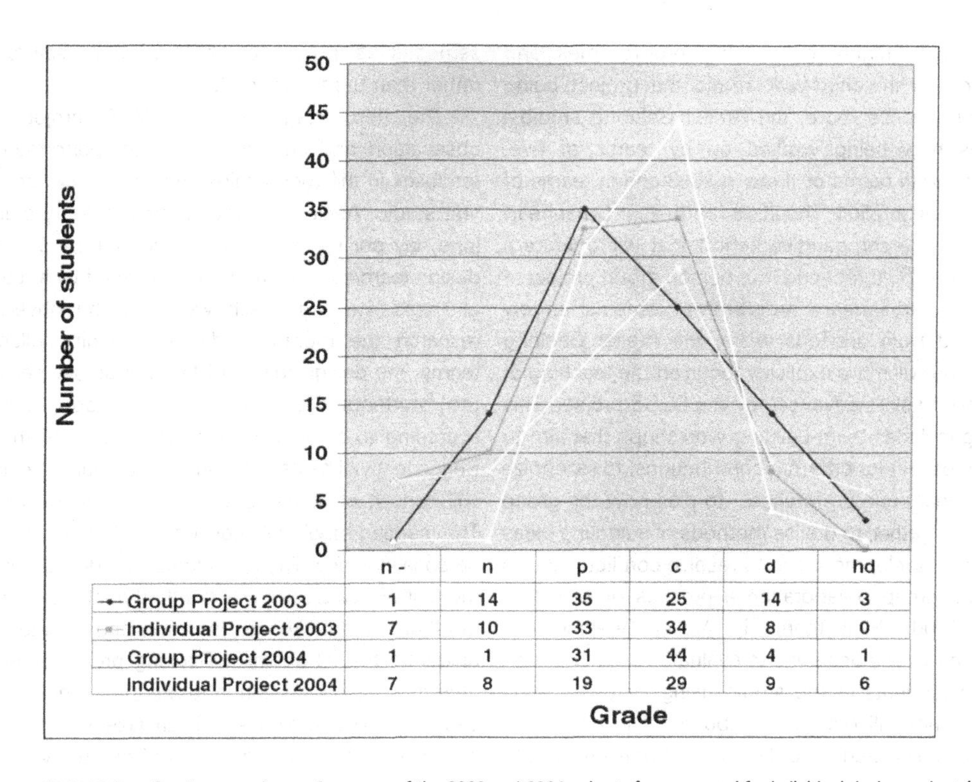

FIGURE 1 Graph comparing performance of the 2003 and 2004 cohorts for group and for individual design projects[1]

It can be seen from Figure 1 that the distribution of grades across the 2004 cohort, in common with the 2003 cohort, does not correspond to the bell curve of normal distribution. For 90% of students, rather than the 68% of a normal distribution, lie within one standard deviation of the mean grade. By uniformly allocating students to teams according to ability (as indicated by their grades in the previous semester), the grades of the groups were effectively evened out. Not one group failed the 2004 project, yet in line with this there was only one higher distinction project. Although it is arguable whether the more even distribution of marks is desirable, the level of marks across the year would indicate that group working is becoming a success, if not altogether a popular one. In 2004, the average mark of 66% for the group project was a notable 8 percentage points higher than the 58% average achieved by the same students for the subsequent individual project. Moreover, a comparison between the performance of this cohort and that of the previous year's leads to further deductions; for whereas the 2004 cohort outperformed the 2003 cohort by only 1.8 percentage points for the second semester second-year design course and by an average of 1.6 percentage points across all other courses, in the Atelier Geelong project, the 2004 students outperformed those of 2003 by 7 percentage points. This might lead to the deduction that the focus on collaboration skills in 2004 increased the effectiveness of the Atelier project as a learning experience.

CONTINUATION INTO 2005

The 2005 STALG-funded group learning project that builds on the first two years of research at Deakin addresses its principal research questions through three forms of evaluation – formative evaluation through questionnaires, summative evaluation through reflective folio assessment and analysis, and illuminative evaluation through focus group discussions, observation of tutorials and analysis of student work.

Two cohorts are being closely observed in 2005 across two year levels taking part in two group design projects with highly contrasting programmes and structures. In the third-year studio the project being observed is, once more, the Atelier Geelong studio – which is now being worked on by teams of five, compared with teams of three in 2003 and in teams of six and seven in 2004. The 2005 fourth-year Urbanheart studio forms a comparative cohort that will operate a number of two-, three- and four-person group projects.[2] While the Urbanheart programme remains largely unchanged from previous years, the Atelier Geelong programme will more explicitly focus on the teaching of teamworking skills advanced by the CUDE project. This teaching includes team-building workshops that aim to develop respect for different contributions, to recognize different team role preferences, to plan how the group will work together, to devise methods of selecting ideas for further development and to resolve conflicts, and to build into group collaboration a process of reflection (Fisher, 2000). Nine teams in Atelier Geelong and Urbanheart will be observed to evaluate communication between students in the studio using an observation template that will record contribution. The nine third-year teams, selected according to their make-up and to voluntary participation, will be observed in three contrasting situations including design meetings which echo those of professional practice, formal sign-up tutorials and informal collaboration in the studio. In contrast, the nine fourth-year teams will be chosen at random. During semester 2 of 2005, control samples in the form of students' performance in design units largely comprised of individually assessed design projects will be analysed for comparison.

The 2005 STALG-funded group learning project can be divided into five stages, which we shall now look at in detail. The first is a survey. This will include the posting, collection and collation of questionnaires from unit chairs relating to student performance in individual versus group design projects from invited schools throughout Australasia and the UK. The second stage is personality type testing which will take place at the beginning of the Atelier Geelong programme when students will be invited to sit the Keirsey Temperament Sorter II (Keirsey, 1998). This is a 70-question psychological characteristic assessment that is based on the Myers Briggs test.[3] The testing of fourth-year students will not take place until the end of the semester as group membership will be self-regulated rather than predetermined.

The third stage of the STALG project is the observation and analysis of student performance and feedback in the two studios previously described. In the first studio, Atelier Geelong, central to every studio day (one day per week) for each team will be a 30-minute design-team/client-team meeting chaired by a tutor. Nine of these meetings each week will be observed. To research the merits of diversity within collaborative teams, the design teams of five will be self-selected by the students from within three pools composed according to personality type. The results of the Myers Briggs test will be used to form these pools. The first pool will consist of students all of similar personality types. The second pool will consist of students of diverse personality types. Those students who do not consent to the test will be placed within a third pool with a selection of other students chosen randomly and will therefore be unidentifiable. The Atelier Geelong project is extended after its final submission in to the three-week landscape design project known as Piazza Geelong. This design exercise is undertaken in teams of one to five (who and how many are in each team is determined by the students) and introduces a major and unexpected shift in the brief that challenges teams to respond to the kind of design crisis that is common in professional practice. The shift, which is usually at first vehemently resisted, asks students to develop not, as they might expect, their own designs but instead those of their colleagues. In the second studio, Urbanheart, a series of parallel urban design projects are run. While the studio traditionally operates on a three-member design team, there will be two-, three- and four-member groups established in 2005 in order to test for the optimum group size for collaborative-based learning and assessment.

The fourth stage of the STALG project consists of two audio-taped focus group discussions that will aim to provide illumination of issues raised in the questionnaires and in the observation of tutorials. 10 students will be randomly selected for the first discussion group, and for the second there will be two groups each of 10 invited students. The first of these groups shall consist of the same students that took part in the first discussion; the second group will be selected to represent diversity. The fifth and final stage of the project consists of a digital folio and unit assessment evaluation and moderation. Unit coordinators at Deakin must complete a detailed study of students' folios to ascertain final grades. In order to distance the results of this assessment from the findings of the STALG project, a number of folios selected randomly from three students per marking category will be moderated by a design tutor not involved in the teaching of the two units.[4]

The first design project of 2005 – a one-day team esquisse focusing on brainstorming and idea selection – allowed students to self-select teammates in order to eventually determine if personality type might influence students' choice of teammates. The exercise has revealed a significant if not unexpected determinant of self-selected team structure – collaborative history. Prior to the third year, there had been three team projects set in three different units, and 74 of the 86 students taking part in the first team esquisse of the third year had been involved in at least one of these projects. Of these 74 students, 48 chose to work with someone whom they had worked with on one of the three previous collaborative projects. And hence it would appear that for 65% of the 2005 third-year design cohort, the experience of having worked before with someone was a significant influence on their choice of teammate. Our further research would hope to reveal what determines the choice of teammate when there is o collaborative history.

The 2005 Atelier Geelong cohort of 95 students have, at the time of writing this paper, been introduced to the STALG project and invited to take part in personality type testing and studio observation. 76 of the 95 students were present at the introduction. All 76 students signed Plain Language Statements agreeing to take part in the observation and 67 completed the Keirsey personality test.

Among the project team there has been a degree of deliberation over how diverse the personality composition of an architectural cohort could be. The results of the type testing have confirmed what the experience of the design teachers at Deakin had suggested – namely that architectural students are, as illustrated in Figure 2, of a wide range of personality types, but that also a number of 'function'[5] types dominate their motivation. At this stage of the STALG project there are a number of results from the test that are worthy of comment. Of the 67 students that have thus far sat the test (30 were female, 37 male), 57% can be described as extroverts and 43% introverts. There was an equal 50/50 split between extrovert and introvert in the females, whereas 62% of the males were extrovert. The most common of Jung's eight types in the cohort is the 'extroverted sensation' type, which number 26 (39% of those tested, of which 61% were male), while the least common was the 'introverted thinking' type, which numbered only six (9%). It might be

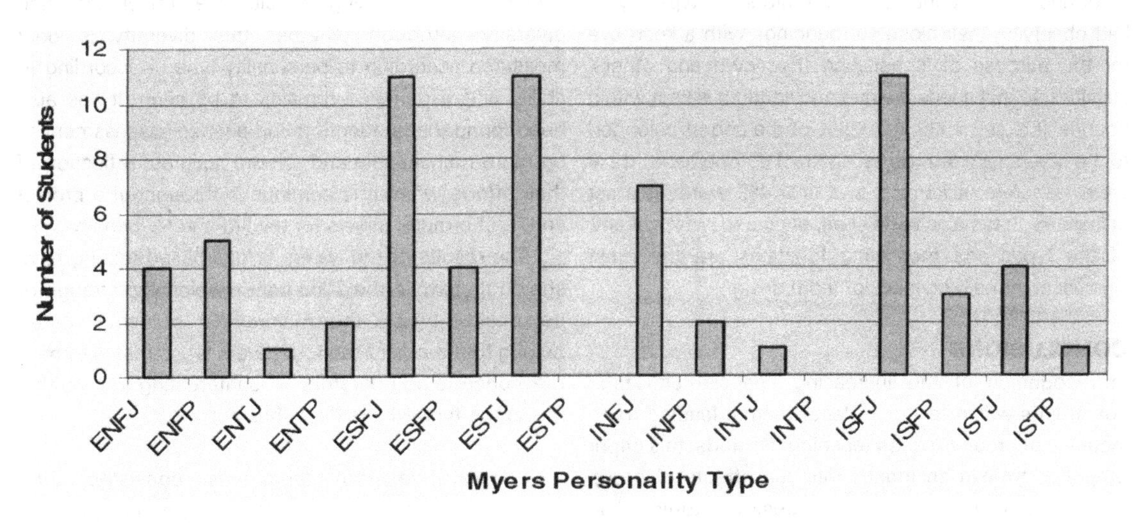

FIGURE 2 Graph showing composition of 2005 cohort as determined by Keirsey Temperament Sorter

interesting to quote Jung's thoughts on what is the most common type among our group of student architects:

> *No other human type can equal the extraverted sensation type in realism. His sense for objective facts is extraordinarily developed. His life is an accumulation of actual experience with concrete objects, and the more pronounced he is, the less use does he make of his experience...Such a type – the majority are men apparently – does not, of course, believe himself to be 'subject' to sensation. He would be much more inclined to ridicule this view as altogether inconclusive, since, from his standpoint, sensation is the concrete manifestation of life – it is simply the fullness of actual living...To sense the object, to have and if possible to enjoy sensations, is his constant motive...His love is incontestably rooted in the manifest attractions of the object...his fastidious taste is obliged to impose certain claims upon his entourage. He even convinces one that certain sacrifices are decidedly worth while for the sake of style. (Jung, 1921)*

67% of the cohort was driven by 'sensation' rather than 'intuition', and 70% by 'feeling' rather than 'thinking'. Moreover, and perhaps most notably, 90% could be characterized as 'judging' rather than 'perceiving'. Thus, 55% of the cohort conform to one of the four Myers type is termed as the 'concrete co-operators' – a type Myers had observing their close surroundings with a keen eye for the purpose of 'scheduling their own and others' activities so that needs are met and conduct is kept within bounds' (Keirsey, 1998: 19). 25% of the cohort belonged to the Myers type termed as 'abstract co-operators', 12% were 'concrete utilitarians' and only 4% were 'abstract utilitarians'. It remains to be seen, of course, which, if any, of the types and motivating functions are the most significant in the processes of team design.

CONCLUSIONS

In recognition of the increasing financial and time constraints within teaching departments, leading to an increase of group-focused teaching methods, this paper proposes to lead an inquiry into the effects of group management in the student design studio. The processes by which this assessment will be carried out will centre on group manipulation guided by a Myers Briggs-based personality scale. This scale will be used to manipulate the dependent cohort's group structure and will be used to compare it with the structure of the control group that will be free to self-select their own teammates. It is hoped that this investigation will begin to highlight the strengths and weaknesses of each style of group formation and begin to inform best-practice principles for teaching group design projects.

Research at Deakin thus far strongly suggests that students' performance is appreciatively higher for group design projects compared with individual projects, and that this improved performance is enhanced further when collaborative skills are taught as part of the group projects. A successful andragogy for collaborative design projects, it would seem, must include teaching collaborative teamworking skills and perhaps even introduce into the curriculum project management techniques centred on research models of group formation. Student feedback indicates that collaboration would be a more positive experience if there were less than six in a group and if students had a degree of control over teammate selection and the assessment of their work. Although students would prefer to self-select groups, it has been found that the performance of the cohort as a whole improves when groups are engineered to contain a range of experience and abilities (Schrage, 1995). It is suggested, therefore, that groups should be self-selecting according to rules that encourage their diversity – although whether this diversity is best measured according to personality type or according to ability and experience remains to be seen. It has also been found that students found peer/self-assessment to be more manageable and a more accurate reflection of their efforts when it is continuous throughout a project and, furthermore, allows for reward and for penalty.

The results of the Myers Briggs-based testing have shown that while the 2005 cohort belong to a range of personality types, the majority of students tested belong to the one of the four Myers types that is termed the 'concrete co-operators'. It is interesting to note that this is the type Myers describes as:

> *'Conservative and stable'; as 'consistent' and 'routinized'; as 'sensible', 'factual' and 'unimpulsive'; as 'patient', 'dependable' and 'hard-working'; as*

'detailed', 'painstaking', 'persevering' and 'thorough'. (cited in Keirsey, 1998: 19)

This is described by Keirsey (1998: 19) as a clear-cut pattern of action and attitude that is highly unlike that of the three other types. It is surely also a pattern of action and attitude that would appeal to many prospective clients of architects.

The 2004 Deakin Student Life report concluded (p13):

While staff cannot be completely responsible for group work processes and functioning, given the notion of experiential and adult learning, they can provide students with greater support in the provision of resources.

This sentence highlights what is perhaps the feeling of most people involved in this research project to date. For what has become clear, in line with the findings of the CUDE project, is that students need help to develop the communication skills essential for effective collaboration, and that this help must come from tutors who have the time and moreover the training to carefully teach and even model these skills. In other words, the notion that group projects demand less teaching resources is based on a false economy.

AUTHOR CONTACT DETAILS

Richard Tucker: School of Architecture and Building, Waterfront Campus, Deakin University, 1 Gheringhap Street, Geelong, Victoria 3217, Australia. Tel: +61 3 5227 8308, fax: +61 3 5227 8303, e-mail: rtucker@deakin.edu.au

John Rollo: School of Architecture and Building, Waterfront Campus, Deakin University, 1 Gheringhap Street, Geelong, Victoria 3217, Australia. Tel: +61 3 5227 8329, fax: +61 5227 8303, e-mail: ajrollo@deakin.edu.au

NOTES

1 Where n – is equivalent to less than 35%; n (fail) is 35% to 50%, p (pass) is 50% to 60%, c (credit) is 60% to 70%, d (distinction) is 70% to 80%, hd (higher distinction) is 80% or above.

2 Urbanheart integrates students entering their second degree into a research culture by blending 'atelier' and 'laboratory' models of the studio. It draws from professionally oriented research models, such as clinical research from medicine and the health sciences, where processes of observation, investigation and exploration are conducted in-situ by the 'practitioner as researcher' (Chenail and Maione, 1997). The forum, which in this case is based around an architecture/urban design programme, allows students to engage in critical discourse by working on complex strategic design initiatives in two significant areas – metropolitan urbanism and regional urbanism.

3 The psychological types identified by the Myers Briggs test were first discussed by Carl Jung in his book *Psychological Types*. Jung posited that people have a multitude of instincts, which he termed 'archetypes', that drive them. He argued that our natural inclination to either 'attitude' type – 'extraversion' or 'introversion' – is combined with our preference for one of what he called the 'four basic psychological functions' – namely, 'thinking' (identified by the letter I in Myers Briggs testing), 'feeling' (identified as T), 'sensation' (identified as S) and 'intuition' (identified as N). Our preference for a psychological function is characteristic, he wrote, and so we can be identified or typed by this preference. This gave Jung eight psychological types – the extraverted thinking type, the extraverted feeling type, the extraverted sensation type, the extraverted intuitive type, the introverted thinking type, the introverted feeling type, the introverted sensation type and the introverted intuitive type. With the addition of two further characteristics, either judging (J) or perceiving (P), Myers Briggs-based tests identify 16 personality types – ENFJ, ENFP, ENTJ, ENTP, ESFJ, ESFP, ESTJ, ESTP, INFJ, INFP, INTJ, INTP, ISFJ, ISFP, ISTJ and ISTP. Of these 16, Myers describes for broad groups, the 'concrete co-operators' (the SJs), the 'abstract co-operators' (NFs), the 'concrete utilitarians' (SPs) and the 'abstract utilitarians' (NTs).

4 An ethics application for the STALG project has been approved which acknowledges and addresses the primary ethical principles involved in this field of research. These include beneficence and non-maleficence, fidelity and responsibility, integrity and justice.

5 See note 2.

REFERENCES

Anderson, K., 2004, *Group Work in Higher Education: Report, Resources and Recommendations* (unpublished).

Chambers, T., Manning, A. and Theriot, L., 2000, 'A new theory for the assignment of members to engineering design teams', *Proceedings of the ASEE Gulf-Southwest Annual Conference, Las Cruces, New Mexico,* April 5–8, 76–82.

Chang, T. and Chang, D., 2000, 'The role of Myers-Briggs type indicator in electrical engineering education', http://www.ineer.org/Events/ICEE2000/Proceedings/papers/MD6-2.pdf (accessed 10 August 2005).

Chenail, R.J., and Maione, P., 1997, 'Sensemaking in clinical qualitative research', in *The Qualitative Report*, 3(1), http://www.nova.edu/ssss/QR/QR3-1/sense.html (accessed October 2004).

Cuff, D., 1991, *Architecture: The Story of Practice*, Cambridge, Mass, MIT Press.

Deibel, K., 2005, 'Team formation methods for increasing interaction during in-class group work', Source Annual Joint Conference Integrating Technology into Computer Science Education, *Proceedings of the 10th Annual SIGCSE Conference on Innovation and Technology in Computer Science Education, Caparica, Portugal, 27–29 June 2005*, 291–295, http://www.cs.washington.edu/homes/deibel/sigcse2005paper.pdf (accessed 13 September 2005).

Dillenbourg, P., 1999, 'What do you mean by collaborative learning?' in P. Dillenbourg (ed.), *Collaborative-learning: Cognitive and Computational Approaches,* Oxford, Pergamon, 1–16.

Durling, D., Cross, N. and Johnson, J., 1996, 'Personality and learning preferences of students in design and design-related disciplines', IDATER, http://www.lboro.ac.uk/departments/cd/research/idater/downloads96/durling96.pdf (accessed 13 September 2005).

Ellis, T.J. and Hafner, W., 2005, 'Peer evaluations of collaborative learning experiences conveyed through an asynchronous learning network', in System Sciences, 2005, HICSS apos 05, *Proceedings of the 38th Hawaii International Conference on System Sciences (HICSS-38), Big Island, Hawaii* 2005.

Falchikov, N., 1986, 'Product comparisons and process benefits of collaborative peer group and self-assessment', in *Assessment and Evaluation in Higher Education,* 11(2), 146–166.

Fisher, A., 2000, 'Developing skills with people: a vital part of architectural education' in C. Nicol and S. Pilling (eds), *Changing Architectural Education: Towards a New Professionalism,* London, Spon Press, 141.

Gorla, N. and Lam, Y., 2004, 'Who should work with whom?', in *Communications of the ACM,* 7(5), 79–82.

Grigg, L., Johnston, R. and Milsom, N., 2003, *Emerging Issues for Cross-Disciplinary Research: Conceptual and Empirical Dimensions*, Canberra, Commonwealth Department of Education, Science and Training.

Jung, C.G., 1921, *Psychological Types*, translation by H. Godwyn Baynes (1923), New York, Pantheon Books 458–459.

Keirsey, D., 1998, *Please Understand Me II: Temperament, Character, Intelligence,* Del Mar, CA, Prometheus Nemesis Books.

McGourty, J., Dominick, P., Besterfieldsacre, M., Shuman, L. and Wolfe, H., 2000, 'Improving student learning through the use of multisource assessment and feedback', *ASEE/IEEE Frontiers in Education Conference, Kansas City, 18–21 October 2000*.

Oakley, B., Felder, R., Brent, R. and Elhajj, I., 2004, 'Turning student groups into effective teams', in *Journal of Student Centered Learning*, 2(1), 9–33.

Plato, 1955, The *Republic*, translation by D. Lee, Harmondsworth, Penguin Books.

Sanz-Menendez, L., Bordons, M. and Zulueta, M.A., 2000, 'Interdisciplinarity as a multidimensional concept', in *Research Evaluation,* 10(1).

Rutherfoord, R., 2001, 'Using personality inventories to help form teams for software engineering class projects', *Proceedings of the 6th Annual Conference on Innovation and Technology in Computer Science Education,* Canterbury, 25–27 June 2001. http://delivery.acm.org/10.1145/380000/377486/p73-rutherfoord.pdf?key1=377486&key2=2836563211&coll=GUIDE&dl=GUIDE&CFID=50404268&CFTOKEN=32739954 (accessed 10 August 2005).

Schrage, M., 1995, *No More Teams! Mastering the Dynamics of Creative Collaboration*, New York, Currency Doubleday.

School of Architecture and Civil Engineering, 2004, University of Bath, course description, http://www.bath.ac.uk

Smith, E. and Mackie, D., 2000, *Social Psychology,* Santa Barbara, CA, Psychology Press, 94–5.

ARTICLE

Engaging Learners

The Development of Effective E-learning Applications for Students of the Built Environment

David L. Dowdle

Abstract
e-learning does not appear to have been as widely adopted by academics within the built environment discipline as one might have expected. The literature review and research presented in this paper put forward an argument for the introduction of e-learning within built environment programmes taught both in the UK and internationally. This paper presents findings from a comprehensive literature review to revisit exactly what learning means and examines the activities required to achieve effective learning via disparate delivery mechanisms. The benefits of multimedia rich e-learning are highlighted and a guidance and audit tool for achieving 'effective' e-learning is presented for use by built environment lecturers and learning technologists. The tool has been designed to encourage and enable academics to develop/introduce e-learning to their teaching. A small research project is presented in which an e-learning application was presented to students for evaluation and reflection. The results suggest that students are very keen to be exposed to e-learning in a 'blended learning' environment.

■ *Keywords* – **e-learning; effective learning; ADDIE; motivation; feedback; multimedia; design guidance; blended learning**

INTRODUCTION

With e-learning now high on the agenda of the UK Government and of all educational sectors, it is clear that e-learning is here to stay. (HEA, 2004)

Though live, instructor-facilitated, face-to-face classroom instruction will not likely be replaced by e-learning, rest assured e-learning is here to stay. (Troha, 2002)

There has been a great deal of hype surrounding the introduction of e-learning tools and applications into higher education. Although e-learning has come a long way in recent years, it has still yet to gain a firm foothold in many areas of education. Those institutions with a strong association with distance-learning programmes have led the way by developing what was then termed 'online learning' and these have now been further developed under the all-inclusive umbrella of 'e-learning'. Other institutions appear to have paid little attention to e-learning and have stuck with traditional face-to-face lecture/tutorial delivery methods or investigated other methods of learning such as 'problem based' or 'student centred'. However, even these institutions have often adopted an Internet- or intranet-based student interface if only to upload the school handbook, timetables and lecture notes. Nevertheless, examples of good e-learning initiatives do exist although they are hard to find in the architectural, engineering and construction disciplines.

This paper is in three parts. The first part examines what is needed in order to create an effective learning environment. It does this by carrying out a detailed literature review in an attempt to identify those activities and environments that promote effective learning via disparate delivery mechanisms. The second part examines the benefits of e-learning and provides details of a guidance and audit tool developed as part of the deliverables agreed for a special interest group (SIG) funded by the Centre for Education in the Built Environment (CEBE) – a centre within the UK Higher Education Academy. The tool has been developed to encourage good practice in the design and development of effective e-learning applications in the built environment.

The third part presents a report on the initial findings from a small two-stage project within the School of Construction and Property Management at Salford University in which students have been evaluating an e-learning application designed prior to the good practice guidelines now developed. The second stage will involve the redesign of the application using the guidance and audit tool for guidance followed by a re-evaluation. Note that the terms 'student' and 'learner' are synonymous in the context of this paper.

THE EFFECTIVE LEARNER

A great deal of research has been carried out investigating those criteria that contribute towards effective learning. Wilkinson (2002) suggests that optimal learning occurs when 'simulation of real world, problem-based activities' takes place. These activities need to occur within a 'safe' learning environment where students can become immersed within, and interact with, learning events in the knowledge that mistakes are not only expected but will deepen their learning experiences.

Wilkinson goes on to suggest several key criteria for effective learning, including:

- learners being 'scaffolded' (a metaphor for the clues, reminders, coaching and other aids used by tutors to support students in mastering a task or concept) as they develop 'self-efficacy' in enabling technologies
- providing continuous feedback
- cognitive loading is eased until basic or 'foundation' understanding is established, cognitive dissonance and challenge being then increased until the complexity mirrors the real world
- pace of learning being controlled by the learner
- allowing peers to collaborate and learn from each other as they learn to work towards common goals
- providing mentors to coach individuals or groups to help remove barriers to understanding, provide guidance and facilitate them in the construction and validation of their learning
- ensuring learning is fun. (adapted from Wilkinson, 2002)

Although the above criteria were developed from an e-learning perspective, they can equally well be applied to any type of learning situation with very little alteration. Furthermore, Biggs (2003) suggests there are four factors that can help create 'good teaching/learning environments', namely:

- a well-structured knowledge base
- an appropriate motivational context
- learner activity, including interaction with others, and
- self-monitoring.

Hence fun, motivation and feedback are seen as key factors within an effective learning environment, as enjoyment often leads to motivation and a motivated learner will almost always find a way to circumnavigate the shortcomings of any learning delivery system (be it face-to-face or any other approach). Regular feedback within the learning environment can ensure learners are aware of how they are performing, in which areas they are excelling and those areas where extra effort is required (McKeachie, 2002).

Motivation is usually a prerequisite for most human activity and there are a number of theories that attempt to define it. One such theory is termed 'expectancy theory' (Vroom, 1964) and offers the following equation:

Motivation =
Perceived probability of success (expectancy)
x
Connection of success and reward (instrumentality)
x
Value of obtaining goal (value)

If a learner does not believe they can be successful at a task, if they do not see a connection between their activity and success or if they do not value the results of success, then there is a strong probability that the learner will not engage in the required learning activity. These three variables must be present and the learner must apply high weightings to them in order for motivation and the resulting behaviour to be positive and effective.

Learners can be either self-motivated (internal) or be motivated by others or by circumstances outside their direct control (external). Internal motivation is the preferred kind of motivation since learners are doing something because they want to, not merely because they are obliged to. If this can be achieved within a particular learning environment, the remaining factors for effective learning become far easier to achieve.

A literature review carried out by de la Harpe *et al* (1999) identified eight key criteria to help ensure effective learning – effective learners:

- have clear learning goals (outcomes)
- have a wide repertoire of learning strategies and know when to use them
- use available resources effectively
- know about their own strengths and weaknesses
- understand the learning process
- deal appropriately with their feelings
- take responsibility for their own learning, and
- plan, monitor, evaluate and adapt their learning process.

The need for clear learning outcomes is not a new idea. Gagne (1985) outlined research that investigated the foundations of effective instruction or conditions of learning. This research led to the classification of types of learning outcome. These were defined by asking 'How can learning be demonstrated?' Gagne favoured avoidance of the use of verbs such as 'know', 'understand' and 'appreciate' and instead preferred the use of straightforward action verbs such as 'state', 'describe' and 'explain' (Bostock, 1996).

Recently, however, educational theorists have cast doubt on the prescriptive, systematic, lecturer-driven approach to learning, i.e. instructivism offered by Gagne, and have championed a constructivist approach instead. In this approach, the lecturer is there to facilitate student learning by helping them to associate new information and skills with prior knowledge and experience and thus construct their own understanding. Learning should therefore be student-driven rather than instructor or lecturer-driven.

Universities often adopt both approaches, instructivism for year-one students, new to the topic and university life, with constructivist approaches being progressively introduced as students become more familiar with the subject material and are more able to make intelligent decisions within their learning environment (Mergel, 1998). However, any approach to instruction or learning should include learning outcomes that are clear and demonstrable by the learner.

'Learning strategies' refers to methods that students use to learn. These strategies are numerous and range from techniques to improve memory to better studying or exam-taking strategies. For example, a strategy for reading that is frequently expounded in study skills texts is the SQ3R strategy (Rowntree, 1998). SQ3R stands for the initial letters of five aspects of studying any printed media, namely 'survey, question, read, recall and review'. Mind mapping, effective note-taking techniques, essay-writing techniques, active listening skills, research skills, effective literature searches, public speaking and presentation skills are just a few of the many strategies that, if employed on a regular basis, allow students to become more effective learners.

To be an effective learner, a student needs to be fully aware of the learning resources available within their learning environment. The learning environment in this context is not just that offered within the confines of the university campus, but also covers the home and workplace environments available to the learner. These resources tend to include library facilities, standard and specialist software and hardware tools, communication tools, survey instrumentation and, often overlooked by learners, fellow students, family, work colleagues and tutors.

If a student is to develop into an effective learner then they must develop an awareness of their own strengths and weaknesses. No one individual tends to be outstanding in all areas of a given discipline but, by playing to their strengths and developing some of their weaknesses, students can enhance their overall performance. Typical self-analysis approaches have been termed 'self evaluation', 'self assessment' or 'skills audits' to name just a few (Cottrell, 2003). Diagnostic tests prior to commencing a module/course are often instigated by lecturers in order to give students an insight into their current knowledge and understanding of a subject. These tests are easily adapted for online deployment.

A great deal of research has been carried out into how we learn and learning styles. One tool for determining learning style that is often cited is the 'learning styles inventory', developed by Kolb (1984). Such tools attempt to introduce students to their preferred learning style(s). Kolb has long been a strong advocator of experiential learning and proposed a four-stage learning process model (Figure 1).

The model can be started from any of the four stages and can be considered a continuous process in that there is no limit to the number of cycles you can move through in a learning situation. A key aspect of this theory is that the reflection stage is crucial; otherwise learners would in fact not learn and would simply continue to repeat their mistakes. Kolb's research found that people learn sequentially, in four stages, with the likelihood being that one stage or mode of learning would become preferred above the other three by the majority of learners and thus developed more than the others. Kolb's experiential learning cycle model identifies that learning is through concrete experience, through observation and reflection, through abstract conceptualization and through active experimentation.

Kolb further suggested that individual learners begin with their preferred style in the experiential learning cycle and that, for effective learning, learners should then progress through the other stages following a clockwise (or anti-clockwise) cycle. Lashley (1995) simplifies the terms used in Kolb's model as 'feeling', 'watching', 'thinking' and 'doing' in an attempt to make it more user friendly.

If students are to be given the opportunity to develop and become effective learners they need to be content with their learning environment and deal appropriately with their feelings, be it at university, work or home. Many things can occur that threaten students' learning environments and, if these threats persist, then learning can be severely inhibited. Cottrell (2003) suggests these threats range from physical ones such as tiredness or hunger; environmental ones, such as too warm, too cold or too noisy; personal ones, such as relationship difficulties and other difficulties, such as money problems and accommodation difficulties, etc. Fear of failure in impending examinations is a well-documented cause of stress in students (Hughes, 2005). Therefore, the ability to make mistakes in a safe, stress-free and supportive environment is one benefit of e-learning that will be examined later.

Many students find it difficult to get all their work done on time. This is particularly true of those new to university life. At school or work, others may help to plan work and 'chase' you to do things but at university this is not the case. While lecturers do set assignments

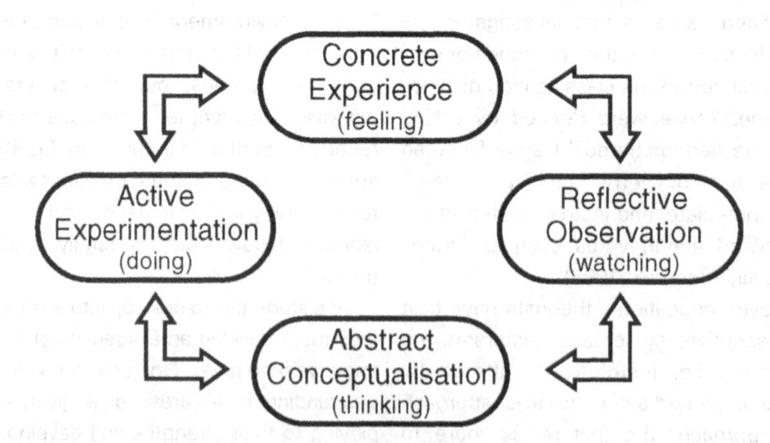

FIGURE 1 Kolb's experiential learning cycle (adapted from Lashley, 1995)

and provide deadlines, they do not normally assist students in the management of their activities and learning. This is something students must learn to do themselves. Time management skills and workload planning are key factors here and need to be adopted quickly if students are to be successful. However, such are the difficulties faced by first-year students, the ensuing panic caused by hand-in deadlines, approaching examinations and failure in these assessments may be one reason for the high drop-out rates at level one. These drop-out rates are not insignificant (HEFCE, 2001). Many online virtual learning environments (VLEs) now provide calendars and other tools to assist students with work planning to help mitigate this problem.

Finally, students need to be encouraged to take more responsibility for their own learning. Student-centred learning (SCL) is one possible direction a school or faculty might take to facilitate students taking more responsibility. SCL is very much learning by doing, learning by practising, making mistakes, reflecting on why those mistakes were made, proposing remedial actions and trying them out. Gibbs (1992) provides a useful definition, stating that student-centred learning, 'gives students greater autonomy and control over choice of subject matter, learning methods and pace of study'. The SLICE (Student centred Learning In Construction Education) project (SLICE, 2002) defines SCL as:

> *Supplementing and occasionally replacing traditional lectures with activities or materials providing some flexibility in terms of the place, pace time and content of student learning.*

These definitions clearly imply that there is a requirement placed on learners to take on a high level of responsibility in their learning. They must actively manage their learning and should not rely on instructors 'spoon feeding' them by telling them what, how, when and where to think. Hogan (1996) cites a need for teachers to hand over more responsibility to their students:

> *I was struck by the irony that I did an enormous amount of reading and thinking about education in order to prepare my lectures, plan effective workshops and select readings and texts for my students, while the students did relatively little. I was the most active learner in my class – because I had total responsibility for what was learned and how it was presented for consumption. (Hogan, 1996)*

De La Harpe *et al*'s (1999) eight key criteria for effective learning were adapted by Dowdle *et al* (2003) when presenting their learning wheel model. When feedback and motivation are included in the learning wheel model as ever-present and all-pervasive key factors (Figure 2), then it is suggested that full recognition and integration of all eight factors when developing a learning environment can significantly help in the creation of effective learners. It is further suggested that the lack of one or more of these criteria will inhibit the chances of students becoming effective learners – analogous to removing the spokes of the wheel or deflating the tyre – making the learning journey prone to delays and frustration.

E-LEARNING

Effectively designed and effectively delivered 'learning events' can take place in any environment, be it a lecture room, tutorial group, laboratory, work, home or on a personal computer (PC). Face-to-face delivery is still the predominant mechanism for taught programmes in universities and colleges. It may have its problems in terms of student attendance, engagement rates, large class sizes and constraining delivery facilities, but there is still a strong argument to keep lectures. Research suggests that students benefit from the support, reassurance and motivation that a tutor can offer; indeed this may be the difference between success and failure for many individuals (McKeachie, 2002; Biggs, 2003).

Good instructors tend to be capable of engaging and enthusing students, pointing them in the right direction, making them more aware and better equipped to develop their own learning styles and strategies. The good lecture is the one where students interact with each other and the lecturer; where they discuss, debate, criticize and listen in turn and not where they dutifully copy notes from the overhead projector or listen to the lecturer without pause for reflection and debate. Good-quality lectures prepare the student for the next stage where they take responsibility for their own learning.

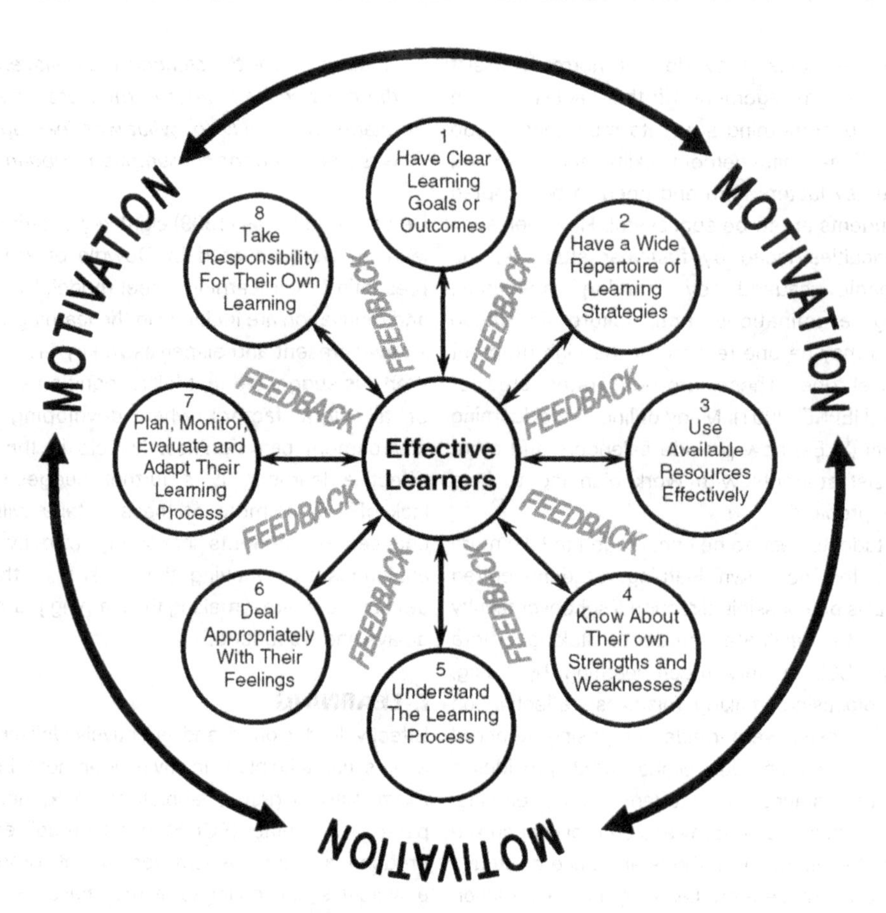

FIGURE 2 **Characteristics of effective learners**

The core message of this paper is that instructors should consider changing the way they impart knowledge and understanding to learners. If the support infrastructure outlined in Figure 2 is available, then the key to creating students who can become effective learners is to employ SCL principles that give students responsibility for their own learning. One way of achieving this is through e-learning.

THE DEVELOPMENT OF E-LEARNING

Good quality e-learning opens up a whole new world of opportunities for lecturers, learning technologists and, most importantly, learners. Not only does it provide new avenues to take in the design and delivery of learning, but also the learning experience created for (and by) the learner can be significant. E-learning should ideally be developed from a clean slate, it is not just a case of taking old course content and delivering it in another mode, but more a case of developing new content based on the needs of the learner and taking advantage of the multimedia capabilities of information and communication technologies (ICT).

A question often asked is 'Why is there a need for a "multimedia" approach to content design – why not stick with just text and a few pictures?' The multimedia capabilities of e-learning applications are viewed as a significant advantage over traditional delivery methods. The ability of multimedia-based learning content to allow the human mind to process information via both verbal and visual processing channels has been demonstrated as more effective than via one channel alone (Mayer, 2001).

Furthermore, Najjar (1996) states that specific situations in which multimedia information may help people to learn include:

- when the media encourage dual coding of information
- when the media support one another, and
- when the media are presented to learners with low prior knowledge or aptitude in the domain being learned. (Najjar, 1996)

This last point suggests that new, level-one university students would benefit from multimedia-rich e-learning applications. Further benefits of multimedia and e-learning in general include significant levels of user interaction, the provision of immediate feedback to the learner, 24/7 access, the ability to repeat sections as often as one wishes and the ability for learners to dictate the pace and location of their learning. Several research studies have shown that good quality, interactive e-learning has many benefits. In a review of numerous meta-analysis studies, Najjar (1996) found that 'learning was higher when information was presented via computer-based multimedia systems than traditional classroom lectures'.

Table 1 provides a list of generally accepted benefits when employing multimedia-rich e-learning (Hick, 1997; CDTL, 2004). If these benefits are capable of being achieved it does seem that e-learning has real potential to improve student performance.

EFFECTIVE E-LEARNING

High quality e-learning applications are not necessarily difficult to create; they do, however, often involve high upfront costs, a dedicated development team and an agreed approach to design in terms of visual interface, navigation rules, media content, learning model, pedagogy, etc.

Good instructional design is often based on pedagogic models such as Gagne *et al*'s (1992) 'nine events of instruction':

- gaining attention (reception)
- informing learners of the objective (expectancy)
- stimulating recall of prior learning (retrieval)
- presenting the stimulus (selective perception)
- providing learning guidance (semantic encoding)
- eliciting performance (responding)
- providing feedback (reinforcement)
- assessing performance (retrieval)
- enhancing retention and transfer (generalization). (Gagne *et al*, 1992)

These nine events can provide the necessary conditions for learning and serve as a basis for instructional design methodologies. If Gagne *et al*'s (1992) model of instruction is to be adopted, then surely the first event, gaining attention, is key.

Learners need to be motivated from the outset of the learning activity to ensure effective learning occurs.

TABLE 1 Benefits of e-learning (adapted from Hick, 1997; CDTL, 2004)

Benefits
Presents learning content in more than one media type – multimedia. Text, graphics, audio, animation and video
Provides opportunity for students to take responsibility for their learning
Capability to be highly interactive, exploratory learning, simulations, quizzes, etc.
Engaging – can be fun! Exciting and challenging. Enhances student motivation
Flexible access to content. 24/7 availability at home, university, work
Non-linear learning sequence – learners pick any route through the content they desire
Can adapt to multiple learning styles
Supports self-paced learning – the Martini effect: anytime, any place, anywhere
Repeatable and non-threatening – if at first you don't succeed, you can try again and again without embarrassment or ridicule
Modular – learners can skip topics they are competent at to concentrate on those they need. Diagnostic tests can help here
Practical – real life simulations, learning by doing
Consistent – all learners cover the same principles and skills. ICT tends to 'force' developers to better organize learning materials
Timely – Just-in-time facility, learners access content when they really need it
Engages multisensory learning modes
Cost effective – high development costs but can prove cheaper long term if content has a relatively long shelf life.

TABLE 2 Attention getting/holding devices from least to most effective (Taylor, 1992 cited in Alber, 1996)

1	Text only
2	Static visual/graphic (varying by colour and size)
3	Animated visual/graphic (silent)
4	Sound alone
5	Sound and movement

If the learning activity is to be effective in gaining learners' attention, then it is important that the best method of activity/learner communication is employed. Research by Taylor (1992, cited in Alber, 1996) suggests that although text can communicate information fairly well, it is not the most effective way to learn. Table 2 lists attention getting/holding devices from least to most effective.

The instructional events described above form a part of a larger approach to learning activity development known as instructional systems design (ISD). ISD is the terminology applied to the 'distinct systematic process through which evolves an instructional project' (Crawford, 2004). Crawford further suggests that 'instructional design delineates a process to follow, through which a conception and understanding of the complex problem is derived'.

ISD is a sequential design methodology in which each step or phase is completed before moving on to the next – the waterfall methodology (Allen, 2003). One approach to e-learning development is via the classic ISD methodology often referred to as the ADDIE method. ADDIE is so described because of its five phases of 'analysis, design, development, implementation and evaluation' (Figure 3).

During the analysis phase, a clear understanding of the needs of the learners is sought. Their desired behaviours or outcomes are considered and the gaps between these and the learners' existing knowledge and skills are defined. In the design phase, key aims and objectives (learning outcomes) are specified together with content, exercises/activities and methods of assessment. Learning materials are then adapted or created within the development phase and then delivered to (or used by) learners in the implementation stage. After implementation is complete, the learners (and the e-learning application) are evaluated to

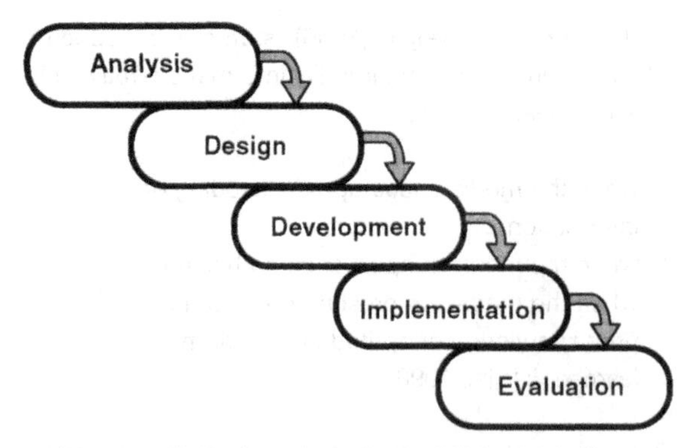

FIGURE 3 The ADDIE method of instructional design

determine the effectiveness of the overall project. Each following phase builds upon the deliverables achieved by the preceding one, without going back to redevelop previous phases based on the findings and feedback obtained in the succeeding phases. This has led to ISD being criticized in many quarters as being too linear in its approach, too inflexible and too constraining in its use. Additionally because each phase must be completed and 'signed off' by the client before moving on to the next one, it can be very time consuming.

An alternative ISD model (Allen, 2003) that has found favour in recent years suggests that rather than developing the instructional project in phases, design teams should take a holistic view while developing the project iteratively. Using this mandate, the design team can work together throughout the project and quickly build exemplar modules (rapid prototyping), which are tested with a learner group and then redesigned based on their feedback. The ADDIE model then becomes a cyclical, repetitive process in which designs grow from an initial seed idea into a final product through a spiralling evolutionary process. Allen (2003) suggests that this iterative approach to ISD (termed 'successive approximation') significantly reduces development time, while also ensuring that all contributors, including learners, have been part of the development process (Figure 4).

A similar iterative approach is suggested by Clark (2004) in which the four ISD stages of analysis, design, development and implementation are operated in parallel with a continuous evaluation process (Figure 5).

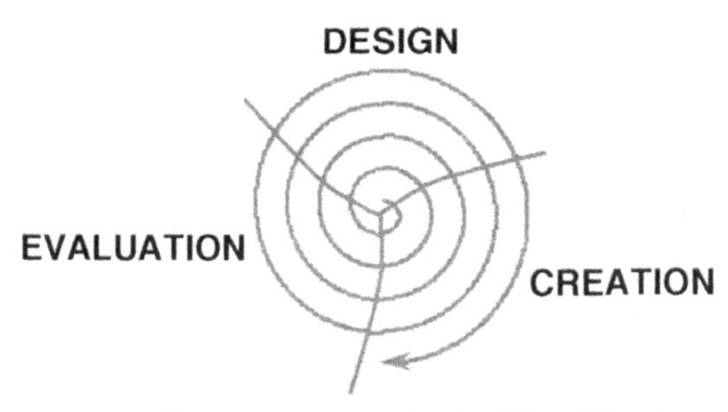

FIGURE 4 The successive approximation ISD model (Allen, 2003)

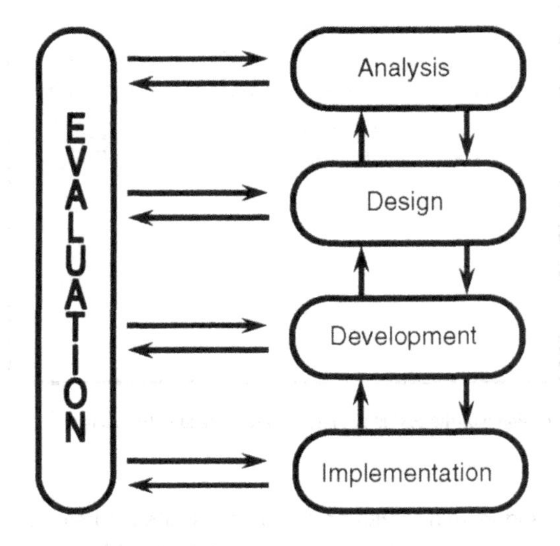

FIGURE 5 The ADDIE rapid prototyping method (adapted from Clark, 2004)

E-LEARNING IN THE BUILT ENVIRONMENT

There are a multitude of sources of instructional design models, guidance and exemplars in effective e-learning both in the UK and internationally. Examples include: M.D. Merrill's 'first principles of instruction' (Merrill, 2002); JISC's 'effective practice with e-learning' guidance and case studies (JISC, 2004); the 'e-learning series' produced by the UK LTSN (now Higher Education Academy) generic centre (LTSN, 2003); 'elements of effective e-learning design' (Brown and Voltz, 2005); and MIT's 'usability guidelines' (MIT, 2004).

Using these and other resources, a tool for enabling high quality e-learning design in built environment disciplines is currently being developed. The development is part of the remit of a Special Interest Group (SIG) sponsored by the UK Centre for Education in the Built Environment (CEBE); a subject centre within the UK Higher Education Academy. The tool, in the guise of an audit form, provides guidance on those aspects that contribute to high quality, effective e-learning. The audit is broken up into the following sections – general details, application introduction, learning outcomes, navigation, usability, system/user feedback, multimedia content, interactivity, assessment, pedagogy/instructional systems design and accessibilty.

Each section provides guide notes as well as a complementary check-off list and Likert-style rating tool. As an example of content, the following text is contained within the 'Guide notes for pedagogy/instructional systems design (ISD)' section of the audit form:

- Generally, words and pictures are better than words alone.
- Corresponding words and pictures should be placed near each other on a page or screen.
- Corresponding animation and narration should be presented simultaneously.
- To avoid overloading the visual channel present words as concurrent narration rather than on-screen text.
- Do not present more verbal material than is relevant.
- Do not present more realistic or detailed visual material than is relevant.
- Do not present the same verbal material both as narration and as on-screen text.
- Use concrete, every day examples to help learners understand abstract concepts and principles.
- Forms of multimedia representations should be appropriate for the instructional situation. For example, use animation for representing dynamic principles.
- Signal the main points to help guide the learner's attention.
- Learners work harder when the presented material is understandable, fits their goals and provides an adequate level of challenge.
- Multimedia instruction has been found more beneficial for learners with low prior knowledge than high prior knowledge.

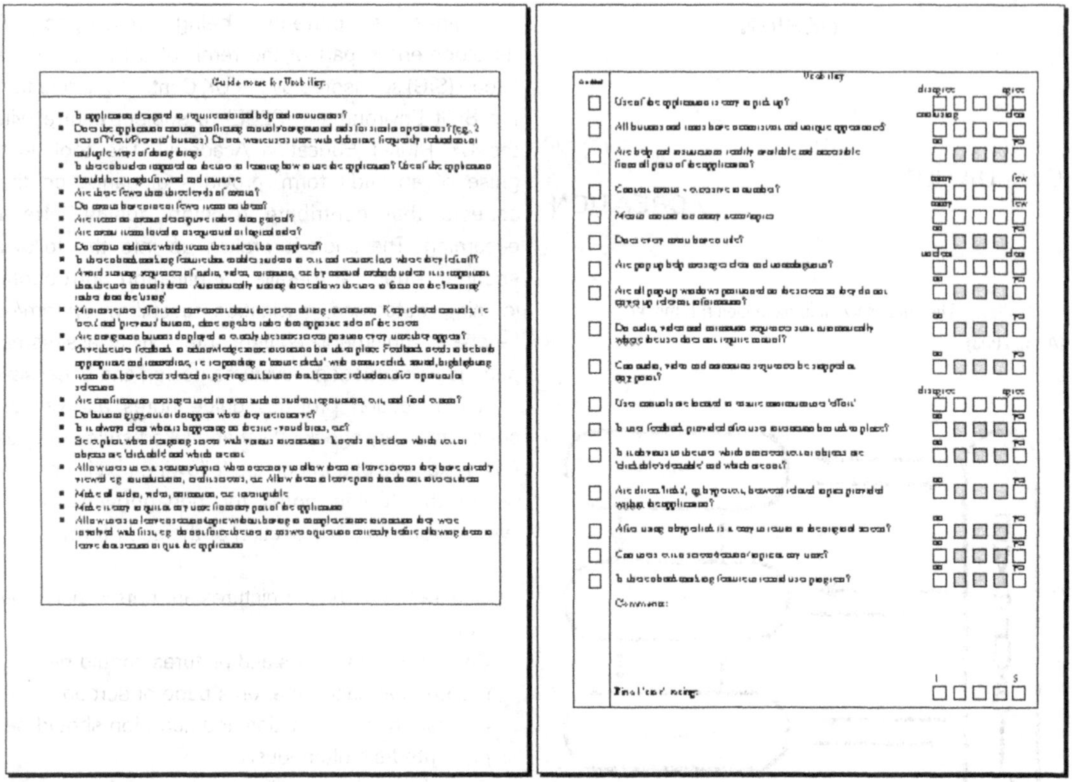

FIGURE 6 Example pages from the CEBE e-learning audit form – version 5.0 (the text is not intended to be read at this scale)

An example image of typical guidance notes and checklist are shown in Figure 6.

To test whether the e-learning design guidance provided in the CEBE audit would result in a positive student learning experience, it was necessary to evaluate an existing e-learning application. An application was available that had passed through the ISD stages of analysis and design without significant levels of evaluation. It was thus ideal for evaluation at this point, as suggested by the rapid prototyping models of ISD, and could therefore be evaluated for e-learning design good practice at the same time.

THE E-LEARNING APPLICATION

The e-learning application in question has been developed by the author in recent years and can be described as an interactive, multimedia-based teaching and learning (T&L) or computer-assisted learning (CAL) tool. The tool was developed using Macromedia's Authorware e-learning software authoring tool (Macromedia, 2005). The highly interactive tool integrates graphics, sound, animation, text and video into a compelling media-rich learning solution. Currently, the application has been developed to provide support for a level-one (year-one) built environment module of studies titled Environmental Science. Module content includes heat transfer, thermal comfort, acoustics and lighting.

The application includes:

- a log-on facility linked to a Microsoft Access database
- text, audio, animation and video-based content
- a bookmark facility that gives students a visual record of their progress from previous uses

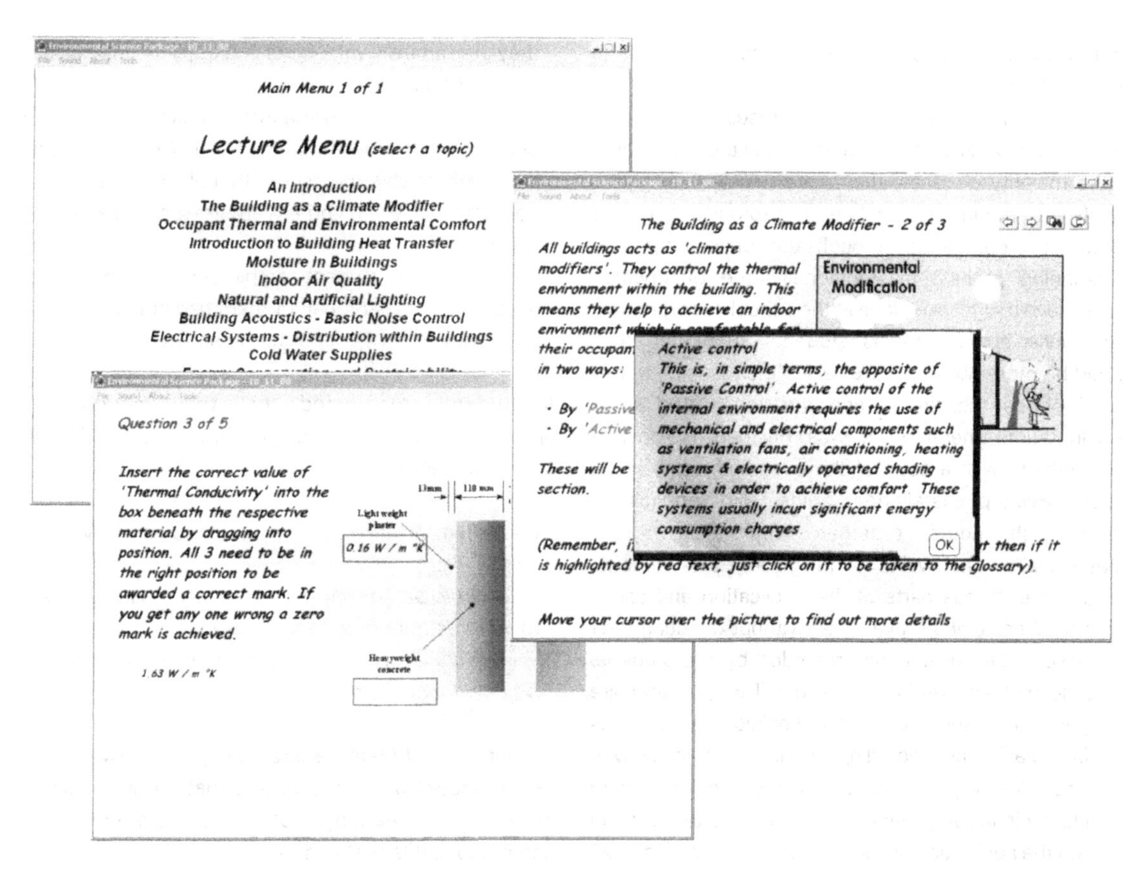

FIGURE 7 A selection of screen images from the sample e-learning application

- a hypertext-based glossary facility allowing students to check the definitions of specific key words
- a note tool that allows students to record text or questions for later recall or submission to the course tutor
- formative self-assessment questions at the end of each subject module that provide instantaneous user feedback
- summative self-assessment questions at the end of the package that provide instantaneous user feedback, recommendations for further study and a link to the Microsoft Access database to allow results to be recorded.

All subject content is contained within the developed executable file. Selecting the executable icon starts the CAL tool running and creates the user database if non-existent, logs on the user, delivers the content, provides regular user interaction and carries out the intermediate and end assessments. Sample images of the application showing the main menu screen, a typical content page with the glossary facility in use and a typical assessment page are illustrated in Figure 7.

METHODOLOGY

A number of research methodologies were considered for the research project including surveys/questionnaires, interviews and observational methods. A multiple method approach of observation followed by structured interview was the initial preferred choice. It was felt that physically watching the actions and behaviours of learners using the e-learning application and then asking them a series of set questions plus questions formulated after observation would allow a deep analysis of learners' individual likes and dislikes to be formulated.

Unfortunately, the timing of the research coincided with a rise in workload for students within the school and, in the majority of cases, they were reluctant to give a significant amount of their free time (predicted as four hours) in order to go through the observation and questioning processes. Additionally, the observation would have had to take place at the school during normal attendance hours whereas students suggested they would be more willing to take part in the research if they could chose the timing and location. After some consideration, therefore, it was decided to provide the students with a copy of the e-learning application, a step-by-step installation guide and application walkthrough notes together with a two-page questionnaire. The walkthrough guided the students through the various parts of the application and asked them to observe and make notes without directing their responses. The observations recorded by the students were clearly their own in this respect. The questionnaire comprised three sections. Section 1 asked four questions requiring basic data regarding learners' familiarity with ICT and e-learning. Section 2 required the learners to provide their initial general observations after working through the application and section 3 asked a series of 26 questions of which 25 used a summated rating (Likert) scale to determine the students' attitude towards the application. The questionnaire was designed by the author of the e-learning application and the questions thus evolved mostly from his developmental experience, although with significant input from literature (Clark and Mayer, 2003; MIT, 2004).

SUMMARY AND ANALYSIS OF FINDINGS

The questionnaire is provided in Appendix 1. Table 3 provides a summary of the results from the 30 questions asked. From this quantitative summary, a number of general observations became clear which were backed up by the qualitative data provided by the students.

GENERAL BACKGROUND OF STUDENT SAMPLE

- All students had been using a PC for at least four years and hence it may be reasonably assumed that they are familiar with accessing applications installed on a PC.
- All students had used the Internet although two had been using it for less than one year. Nevertheless it is felt reasonable to assume that all students are now familiar with using a web browser to search the Internet.
- A quarter of the student sample did not have Internet access at home and half of them had not attempted an e-learning course prior to the survey.

All students responded that they thought the application should be developed further (Q 29). Typical comments included:

Can return to unclear subjects/topics as many times as you like.

I would personally like to use this facility as an aid to lectures but not to replace lectures.

Generally a very good application, I think it should be developed further.

A majority of students agreed that the software was easy to access and start up and that the introduction and expected learning outcomes were clear and unambiguous (Qs 5, 6 and 7).

While there was no very negative feedback on navigation/usability, it was not rated as highly as most other aspects (this was expected) and did result in several comments (Qs 8, 9 and 10):

Automatic skip to next page should be turned off – as this does not allow student time to make own notes if needed.

Movement from page to page was not great i.e. I couldn't move back to the previous page, as it would go back to the first page of the section.

Moving through pages would be better if you didn't have to go through all of them to get back to the menu or a certain point.

The feedback mechanisms provided within the content and the self-assessment sections were favourably received (Qs 11, 12, 13 and 20) although it was surprising to note that 25% of respondents to Q13 did not feel that the provision of feedback was of great

TABLE 3 A summary of results from the e-learning application evaluation

QUESTION	RESPONSES					
1	–		–	2 (12.5%)		14 (87.5%)
2	2 (12.5%)		2 (12.5%)	10 (62.5%)		2 (12.5%)
3		4 (25%)		4 (25%)	8 (50%)	
4		8 (50%)		–	8 (50%)	
	No					Yes
5	–	–	–	–	2 (12.5%)	14 (87.5%)
6	–	–	–	–	1 (6%)	15 (94%)
7	–	–	–	–	6 (37.5%)	10 (62.5%)
8	–	–	–	4 (25%)	4 (25%)	8 (50%)
9	–	–	–	1(6%)	3 (19%)	12 (75%)
10	–	–	–	3 (19%)	7 (44%)	6 (37.5%)
11	–	–	–	–	1 (6%)	15 (94%)
12*	–	–	–	2 (12.5%)	–	13 (81%)
13	–	–	–	4 (25%)	2 (12.5%)	10 (62.5%)
14	–	–	–	–	6 (37.5%)	10 (62.5%)
15	–	–	–	–	6 (37.5%)	10 (62.5%)
16	–	–	–	–	11 (69%)	5 (31%)
17	–	–	1 (6%)	1 (6%)	2 (12.5%)	12 (75%)
18	–	–	–	2 (12.5%)	–	14 (87.5%)
19	–	–	–	3 (19%)	2 (12.5%)	11 (69%)
20 **	–	–	–	2 (12.5%)	4 (25%)	9 (56%)
21	–	–	–	–	6 (37.5%)	10 (62.5%)
22 **	–	–	–	6 (37.5%)	–	9 (56%)
23	–	–	2 (12.5%)	–	6 (37.5%)	8 (50%)
24	–	4 (25%)	2 (12.5%)	4 (25%)	1 (6%)	5 (31%)
25	–	2 (12.5%)	–	8 (50%)	–	6 (37.5%)
26	–	2 (12.5%)	2 (12.5%)	2 (12.5%)	4 (25%)	6 (37.5%)
27	2 (12.5%)	–	–	–	4 (25%)	10 (62.5%)
28	–	–	–	–	4 (25%)	12 (75%)
29	–	–	–	–	–	16 (100%)
30	2 (12.5%)		–	14 (87.5%)		–

* No answer from one student

** Did not work for one student

importance to them. Feedback is regarded as a key element of effective learning (Allen, 2003; Biggs, 2003) and this finding warrants further investigation.

> *Feedback was helpful to those that get the answer right/wrong. For the ones that got it right – it is rewarding; for the ones that got it wrong – it is still reassuring. I consider this interactive feedback to be very important.*

> *Feedback, especially where a wrong answer was picked and the definition for the wrong answer was given – the explanation helped me eliminate that answer.*

The multimedia content was generally well thought of (Qs 14, 15, 23, 24, 25 and 26) and individuals were clearly indicating their media preferences with text-based content being rated highest by the students followed by video, animation and finally audio. While this appears to contradict the findings of past research (Taylor, 1992 cited in Alber, 1996; Clark and Mayer, 2003) it is perhaps understandable. Students were presented with a very incomplete e-learning application that was highly text-based containing only one video and audio element and only two animations. Perhaps the student sample selected would have preferred learning styles not aligned with the visual or auditory aspects of such e-learning applications. Learning styles were not evaluated prior to introducing the students to the application, as there was no time to do so.

However, comments made by the students suggest that had the content been media rich, a higher rating for video, animation and especially audio would very likely have occurred:

Voice recording should be used more as it is very useful. Videos and interactive diagrams excellent.

The lecture on heat transfer is good but I thought there was a little too much text – perhaps to break this up with more use of diagrams?

Maybe there could be different voices to add more variation and interest to the package. The voice should feature throughout most of the lectures.

More sound recordings would help the text 'sink in' as you are reading it. I sometimes found it hard to read from the screen.

The last comment above contradicts good practice (Clark and Mayer, 2003) which suggests that it is not effective to have an audio narration simply repeating onscreen text. Clark and Mayer's redundancy principle suggests that it is best to keep the audio but to remove, or condense to bulleted lists, the text. This is because significant amounts of text may distract the learner from other visual information such as graphics and animation and overload the learners' visual channel by presenting two visual information flows into the brain, which can best operate with only one visual flow at a time (Clark and Mayer, 2003).

The self-assessment facility was highly rated by the students (Qs 18 and 19) and is probably coupled with the instantaneous feedback associated with each of the assessment question types:

I liked the introductory building heat transfer questions – good use of visuals, and good useful feedback.

Multiple choice questions help me learn the best – knowledge but not necessarily understanding.

The other facilities provided by the tool – the hypertext glossary and the notepad – were also rated quite highly although there were some presentation problems (Qs 21 and 22):

I like the way vocabulary is highlighted in red which allows the user to click on it for a definition.

Hypertext links to glossary very useful but text placed directly over main screen text blocking it out.

In fact, placing the glossary text was a deliberate design ploy to ensure that students read the definition before closing the glossary screen to get back to the main screen text. This may need to be reviewed in later developments.

In summary, the e-learning framework and limited content offered to students for evaluation was very well received. Indeed, the positive ratings and feedback were better than expected as the author is well aware of the inefficient and limited capabilities of the application navigation facilities. It would appear that the students coped with these inefficiencies without too many problems, however, this may well be because of the limited amounts of actual content within the application. It is felt that had there been significant amounts of content for the students to work through, the navigation problems would have been much more noticeable to the user.

Perhaps final comments can be left to the students – when asked in the last question of the survey what their preferred method of learning was, 87% stated they preferred a mixture of face-to-face and e-learning i.e. a blended approach:

A combination of e-learning and lectures would help me learn because I could go through things at my own pace and gain basic knowledge. I would then be in a better position to ask questions and further my understanding in lectures.

In my experience, students all have 'off days' where they come to lectures and cannot concentrate.

Fact is, what one misses out in a lecture cannot be regained easily. If this programme can be used by students, they can use it in a relaxed environment and gain more. Coupled with regular face-to-face tutorial sessions, this tool has potential.

CONCLUSIONS AND RECOMMENDATIONS

The literature review undertaken within this research paper has led the author to recommend that all learning environments should strive to meet the eight key criteria that help to ensure effective learning. Effective learners:

- have clear learning goals (outcomes)
- have a wide repertoire of learning strategies and know when to use them
- use available resources effectively
- know about their own strengths and weaknesses
- understand the learning process
- deal appropriately with their feelings
- take responsibility for their own learning, and
- plan, monitor, evaluate and adapt their learning process.

When feedback and motivation are included as ever-present and all-pervasive key elements within these eight criteria (the learning wheel), then it is proposed that effective learners will be created.

The core message of this paper is that we can and must change the way we impart knowledge and understanding to students. If the supporting infrastructure, in terms of student support, learning resources, etc. (the learning wheel), is available to them then the key to creating students who can become effective learners is to employ student-centred learning principles or a similar approach that gives students responsibility for their own learning. This paper recommends that a very effective method of SCL is through e-learning. Good quality e-learning opens up a whole new world of opportunities for lecturers, learning technologists and, most importantly, students. The many benefits of e-learning were shown to include:

- flexible access to content – 24/7 availability at home, university and work
- supports self-paced learning – the Martini effect: anytime, any place, anywhere
- engages multisensory learning modes, and
- timeliness – 'just-in-time' facility, learners access content when they really need it.

Good quality e-learning should follow sound ISD models and good practice guidance in order to achieve acceptance from the learners it is targeted towards. The research highlighted in this paper has identified and illustrated sound ISD models based on a rapid prototyping approach to the traditional ADDIE ISD model. These models, when coupled with the good practice guidance being developed, will help academics and technologists create effective e-learning applications.

To test these proposals, an e-learning application that had been commenced prior to this research by the author was evaluated using a sample of 16 students from the School of Construction and Property Management within Salford University. The results of this evaluation were very positive but did pick up on several problematic aspects of the application. It is felt that these problems would have not occurred if the development had followed the suggested ISD model and good practice guidance audit described earlier.

The key finding, however, is that 87% of the student sample stated they preferred a blended approach to their learning. This last point needs to be given serious consideration by those who are resistant to change in higher education in the UK and internationally.

AUTHOR CONTACT DETAILS

David L. Dowdle: School of Construction and Property Management, University of Salford, UK. Tel: +44 (0) 161 295 2182, fax: +44 (0) 161 295 5011, e-mail: d.l.dowdle@salford.ac.uk

REFERENCES

Alber, A.F., 1996, *Multimedia – a Management Perspective*, Wadsworth, Belmont, US.

Allen, M., 2003, *Michael Allen's Guide to e-learning – Building Interactive, Fun and Effective Learning Programs for Any Company*, John Wiley & Sons, New Jersey.

Biggs, J., 2003, *Teaching for Quality Learning at University*, 2nd edn, SHRE and OU Press, Buckingham.

Bostock, S., 1996, *Instructional Design – Robert Gagné, The Conditions of Learning*, http://www.keele.ac.uk/depts/cs/Stephen_Bostock/docs/atid.htm (accessed 8 January 2005).

Brown, A.R. and Voltz, B.D., 2005, 'Elements of effective e-learning design', in *International Review of research in Open and Distance Learning*, 6(1), March, http://www.irrodl.org/content/v6.1/brown_voltz.html (accessed 23 March 2005).

CDTL, 2004, *Benefits of Multimedia Courseware*, Centre for Development of Teaching and Learning, National University of Singapore, http://www.cdtl .nus.edu.sg/mmi/benefits.htm (accessed 27 September 2004).

Clark. D., 2004, *Instructional Systems Design Concept Map*, http://www. nwlink.com/~donclark/hrd/ahold/isd.html (accessed 27 January 2005).

Clark, R.C. and Mayer, R.E., 2003, *e-learning and the Science of Instruction: Proven Guidelines for Consumers and Designers of Multimedia Learning*, San Francisco, CA, Pfeiffer.

Cottrell, S., 2003, *The Study Skills Handbook*, 2nd edn, Basingstoke, Palgrave Macmillan.

Crawford, C., 2004, 'Non-linear instructional design model: eternal, synergistic design and development', *British Journal of Educational Technology*, 35(4), 413–420, http://www.ingentaconnect.com/search/article;jsessionid= 4q3s3oc530dk2.henrietta?title=ADDIE&title_type=tka&year_from=1997& year_to=20 04&database=1&pageSize=20&index=1 (accessed 22 February 2005).

de la Harpe, B., Kulski, M. and Radloff, A., 1999, 'How best to document the quality of our teaching and our students' learning?' in K. Martin *et al* (eds), *Teaching in the Disciplines/Learning in Context, – Proceedings of the 8th Annual Teaching Learning Forum, The University of Western Australia, February 1999*, Perth, UWA, 108–113, http://lsn.curtin.edu.au/tlf/tlf1999/ delaharpe.html (accessed 15 October 2004).

Dowdle, D., Murray, P. and Parker, M., 2003, 'Student centred learning – the keystone of construction education? in R. Newton *et al* (eds), *Proceedings of the CIB W89: International Conference on Building Education and Research BEAR 2003*, The Lowry, Salford, April 2003, Salford, University of Salford, 542–555.

Gagne, R., 1985, *The Conditions of Learning*, 4th edn, New York, Holt, Rinehart & Winston.

Gagne, R., Briggs, L. and Wager, W., 1992, *Principles of Instructional Design*, 4th edn, Fort Worth, TX, HBJ College Publishers.

Gibbs, G., 1992, *Assessing More Students*, Oxford, Oxford Brookes University.

HEA, 2004, *Thematic Work – e-learning Series*, Higher Education Academy, http://www.heacademy.ac.uk/1682.htm (accessed 18 March 2005).

HEFCE, 2001, *Performance Indicators for Higher Education*, 00/40, HEFCE, http://www.hefce.ac.uk/learning/perfind/2001/download/main_report.doc (accessed 17 October 2004).

Hick, S., 1997, *Benefits of Interactive Multimedia Courseware*, http://www.ftsm.ukm.my/th2723/A83816/BENEFITS%20OF% 20INTERACTIVE%20MULTIMEDIA%20LEARNING.htm (accessed 22 October 2004).

Hogan, C., 1996, 'Getting students to do their reading, think about it and share their ideas and responses', in J. Abbott and L. Willcoxson (eds), *Teaching and Learning Within and Across Disciplines – Proceedings of the 5th Annual Teaching Learning Forum, Murdoch University, February 1996*, Perth, Murdoch University, 79–81, http://lsn.curtin.edu.au/tlf/tlf1996/ hoganca.html (accessed 15 October 2004).

Hughes, B., 2005, 'Study, examinations, and stress: blood pressure assessments in college students', in *Educational Review*, 57(1), 21–36.

JISC, 2004, *Effective Practice with e-learning*, http://www.jisc.ac.uk/ index.cfm?name=elp_practice (accessed 26 February 2005).

Kolb, D.A., 1984, *Experiential Learning: Experience as the Source of Learning and Development*, Englewood Cliffs, NJ, Prentice Hall.

Lashley, C., 1995, *Improving Study Skills – A Competence Approach*, London, Cassell.

LTSN, 2003, *e-learning Series – Guides 1, 2, 3, 4 and 5*, http://www.heacademy.ac.uk/1682.htm (accessed 12 March 2005).

Macromedia, 2005, *Macromedia Authorware 7*, http://www.macromedia. com/software/authorware/ (accessed 22 January 2005).

Mayer, R.E., 2001, *Multi-Media Learning*, Cambridge, Cambridge University Press.

McKeachie, W.J., 2002, *McKeachie's Teaching Tips – Strategies, Research and Theory for College and University Teachers*, Boston, US, Houghton Mifflin Company.

Mergel, B., 1998, *Instructional Design and Learning Theory*, University of Saskatchewan, http://www.usask.ca/education/coursework/802papers/ mergel/brenda.htm#Learning%20Theories%20and%20the%20Pract (accessed 12 December 2004).

Merrill, M.D., 2002, 'First principles of instruction', in *Educational Technology Research and Development*, 50(3), 43–59.

MIT, 2004, *Usability Guidelines*, http://web.mit.edu/ist/usability/usability-guidelines.html. (accessed 18 November 2004).

Najjar, L.J., 1996, 'Multimedia information and learning', in *Journal of Educational Multimedia and Hypermedia*, 5(2), 129–150. www.medvet.umontreal.ca/techno/eta6785/articles/multimedia_and_ learning.PDF (accessed 22 January 2005).

Rowntree, D., 1998, *Learn How to Study – A Realistic* Approach, 4th edn, London, Warner Books.

SLICE, 2002, *Handbook on Student-centred Learning in Construction Education*, Plymouth, University of Plymouth.

Troha, F.J., 2002. *Managing Threats to E-learning Success: Six Simple Tips for Initiative Leaders*, PhD thesis, http://www.e-learningcentre.co.uk/eclipse/ Resources/MANAGING%20THREATS.doc (accessed 18 March 2005).

Vroom, V., 1964, *Work and Motivation*, New York, Wiley.

Wilkinson, D.L., 2002. 'The intersection of learning architecture and instructional design in e-learning', in J.Lohmann and M. Corradini (eds), *Proceedings of the 2002eTee conference, Davos, Switzerland, August 2002*, http://services.bepress. com/eci/etechnologies/33/ (accessed 22 February 2004).

APPENDIX 1 E-LEARNING APPLICATION STUDENT EVALUATION

About me

1. Number of years using a computer (at university, home or work)
   Less than 1 yr ❑ 1–3 yrs ❑
   4–6 yrs ❑ over 6 yrs ❑
2. Number of years using the Internet
   Less than 1 yr ❑ 1–3 yrs ❑
   4–6 yrs ❑ over 6 yrs ❑
3. Do you have Internet access at home?
   No ❑ Yes – dial up ❑
   Yes, broadband ❑
4. Have you attempted an e-learning course previously? (either formally or informally)
   No ❑ Yes, at Salford ❑
   Yes, at work ❑

About the sample Environmental Science e-learning application you evaluated

Please use the space below to record your feedback on the application (prompts: access, learning outcomes, navigation, usability, multimedia, interaction, assessment, feedback)

....
....
....
....
....
....
....
....
....
....
....

Please confirm your level of agreement or disagree-ment with the following statements using the checkboxes provided where checkbox 1 equates to 'strongly disagree' and checkbox 6 equates to 'strongly agree':

	1 2 3 4 5 6
5. The application was easy to access and 'start up'	❑❑❑❑❑❑
6. The initial introduction was very thorough	❑❑❑❑❑❑
7. Learning outcomes were clear and unambiguous	❑❑❑❑❑❑
8. I found it easy to navigate around the e-learning application	❑❑❑❑❑❑
9. The application was 'user friendly'	❑❑❑❑❑❑
10. Moving from page to page and menu to menu was quickly achieved	❑❑❑❑❑❑
11. Feedback was provided when expected	❑❑❑❑❑❑
12. Feedback was informative when provided	❑❑❑❑❑❑
13. The ability to receive instant feedback is important to me	❑❑❑❑❑❑
14. The available learning content was well structured and clearly presented	❑❑❑❑❑❑
15. The mix of text, audio, animation and video enhanced the application (or would enhance the finished application)	❑❑❑❑❑❑
16. The application was very interactive in its design	❑❑❑❑❑❑
17. The interactive elements of the application made the learning experience more interesting	❑❑❑❑❑❑
18. The self-assessment questions helped me/would help me to confirm my understanding	❑❑❑❑❑❑
19. The different types of assessment questions offered created an effective assessment facility	❑❑❑❑❑❑
20. The feedback provided after the final assessment proved/would prove useful	❑❑❑❑❑❑
21. The 'hypertext' glossary facility was/would be helpful	❑❑❑❑❑❑
22. The 'notes' tool was/would be helpful	❑❑❑❑❑❑

23. Text-based content is important for me in understanding the subject matter ❑❑❑❑❑❑
24. Audio-based content is important for me in understanding the subject matter ❑❑❑❑❑❑
25. Animation-based content is important for me in understanding the subject matter ❑❑❑❑❑❑
26. Video-based content is important for me in understanding the subject matter ❑❑❑❑❑❑
27. It is possible to learn this subject using e-learning ❑❑❑❑❑❑
28. Development of complete e-learning applications of a similar design would improve my knowledge and understanding of the subject matter ❑❑❑❑❑❑
29. I believe the e-learning application is worth developing further ❑❑❑❑❑❑
30. My preferred method of learning is:
    Face to face ❑ By e-learning ❑
    A mixture of both ❑ Other means (explain below) ❑

....
....
....
....
....
....

ARTICLE

Supervised Work Experience

The Learning Climate of Construction Companies and the Factors that Influence Student Experience

David J. Lowe

Abstract

The benefits to students of an industrial placement or supervised work experience (SWE) as an integral part of undergraduate degree programmes have long been accepted. Employers use SWE as an opportunity to assess the capability of students prior to offering them permanent employment on completion of their studies. Likewise, students use SWE to review an employer's ability to provide them with the relevant post-graduation experience to enable them to progress to professional qualification. Also, during SWE they assess the construction industry in terms of its working environment and as a long-term career. The findings are presented of an eight-year study into the ability of organizations within the construction industry to provide appropriate learning environments during SWE. Construction organizations are perceived to be supportive in terms of personal and informal support provided by colleagues, and to a lesser extent working practices, but less supportive in terms of the more formal support given by managers, specifically in the use of appraisal systems. Construction organizations need to accurately assess their ability to provide an effective learning environment in order to attract potential employees to a career in construction and to retain students within the industry after SWE.

■ *Keywords* – Construction; learning climate; learning organizations; supervised work experience

INTRODUCTION

The BSc in Commercial Management and Quantity Surveying (CM&QS) is a collaborative degree programme between the University of Manchester (formerly UMIST) and a consortium of the UK's leading construction companies. It aims to educate students in the fields of commercial management and quantity surveying and to prepare them for work in a contracting organization. The ethos of the course is that such education is a blend of academic study and industry-based training.

The inclusion of the longest possible period of industrial training was a specific requirement of the consortium. The programme, therefore, is of four years duration and includes a compulsory year in industry as its third year. Described as supervised work experience (SWE), students are required to complete 46 weeks of paid industrial work experience, which can count towards their experience for the RICS Assessment of Professional Competence. The SWE is not formally assessed for progression to the final year of the programme; however, students are required to submit a critical and detailed report on an aspect of their SWE and present the report to the student group and selected staff as part of their final-year assessment. UMIST and the consortium designed the SWE to include a programme of experience and self-development for the student. Although academics visit the students during the SWE, the sponsoring organization is responsible for providing appropriate industrial training and learning environments.

This paper reports on an eight-year investigation into the learning environment experienced by students during SWE. Subsidiary objectives were to investigate:

- students' perceptions of the learning climate of their organization
- factors that influenced their SWE experience; and
- changes in these perceptions over time (1996-2003).

BACKGROUND

According to Ashworth and Saxton (1992), the purpose of a work placement year is to develop maturity, enable the exploration of the theory-practice link, encourage the development of critical but pragmatic thinking, and facilitate systems thinking. Further, Davies (1990) advocates that the benefits of SWE are 'unique, identifiable and not achieved by other means'. Additional benefits to students include the opportunity to assess the construction industry as a long-term career; assess a potential employer in terms of career prospects, working environment, structured training programmes, enabling them to achieve professional status, etc.; and select a topic for their final-year research project within the context of the employing organization.

A recent report commissioned by the Department for Education and Skills (DfES) concluded that:

- With guidance, students of all ages can learn from their experiences in the world of work to develop their key competencies and skills and enhance their employability.
- Employers value people who have undertaken work experience, been able to reflect upon that experience and then go on to articulate and apply what they have learnt.
- Partnerships between employers and higher education are valuable in promoting work-related learning and in improving the quality and quantity of such experiences (Work Experience Group, 2002).

The Quality Assurance Agency for Higher Education (QAA) has produced a Code of Practice on Placement Learning (Quality Assurance Agency, 2001) which contains eight precepts for the quality assurance of placement learning.

The potential benefits for employers of SWE students include recruiting and training future employees, meeting current labour shortages, expanding a well-prepared labour pool, fostering a positive public image, cultivating business opportunities with universities, and receiving wage subsidies (Jackson and Wirt, 1996).

Construction employers see the inclusion of SWE within degree programmes as essential: it affords them the opportunity to present a wide vision of the potential careers available within the industry (Mann, 2002). Moreover, as a result of the current and projected UK shortage of construction graduates (Dainty and Edwards, 2003) – especially quantity surveying graduates (Cavill, 1999) – employers accept that their priority is to keep students studying on construction courses in the industry when they qualify. To support this, the Construction Industry Council is developing a framework for employers when taking students on work experience (Hampton, 2001), while the CITB launched an industry-sponsored 'Design-a-job' website in 2001.

There are, however, financial implications to be considered by employers providing SWE placements. These include the cost of planning work-based learning; orienting and training staff; training, supervising, mentoring and evaluating students; students' salaries; and poaching (when other companies hire the best students after training) (Jackson and Wirt, 1996).

LEARNING WITHIN THE WORKPLACE

Learning naturally occurs in the work environment (Binsted, 1980) in many situations – formal, informal or incidental. Informal learning within the workplace is predominantly experiential and non-institutional, including self-directed learning, networking, coaching, mentoring, performance planning and trial-and-error (Marsick and Watkins, 1990). Incidental learning, however, is unintentional, a by-product of another activity; examples include learning from mistakes, assumptions, beliefs, attributions and internalized meaning constructions about the actions of others. New experiences or perceptions provide learning opportunities, usually unintended, which may be seized upon or passed over (Rogers, 1986). This natural learning is part of the process of living. In fact, most individual development will occur 'on the job' and not through structured learning activities (Mumford, 1987).

The learning environment at work is more than just a physical area, it contains people and resources – ideas, knowledge and know how. Learners, however, often fail to draw upon the richness of opportunities for learning at work (Harri-Augustein and Thomas, 1991). Also, individuals can be helped or hindered by the organization in which they work, the environment may not be absolutely fundamental but can be a powerful influence (Mumford, 1992).

THE LEARNING ORGANIZATION

The learning organization is a powerful and attractive idea (Salaman, 2001). The term has been defined as 'an organization skilled at creating, acquiring and transferring knowledge, and at modifying its behaviour to reflect new knowledge and insight' (Garvin, 1994); and as 'an organization which facilitates the learning of all its members and continually transforms itself' (Pedler *et al*, 1991). Moreover, the learning organization 'depends absolutely on the skills, approaches and commitment of individuals to their own learning' (Mumford, 1992). However, the learning organization concept is vague, prescriptive and seems to have little foundation in practice (Gardiner, 1999), while Garratt (1999) explains that the learning organization is more an aspiration, 'a vision, which motivates, stretches and leverages the organization for the long term'. Elsewhere, it is seen as a brand (Padaki, 2002).

The term organizational learning, as defined by Tsang (1997), 'is a concept used to describe certain types of activity that take place in an organization, while the learning organization refers to a particular type of organization in and of itself. Nevertheless, there is a simple relationship between the two – a learning organization is one which is good at organizational learning.'

Love *et al* (2000) provide several definitions of both the learning organization and organizational learning. However, to summarize, a successful learning organization should:

- make a commitment to knowledge (Mills and Friesen, 1992)
- have a learning culture (McGill and Slocum, 1993)
- appreciate the significance and dynamics of the learning process (Easterby-Smith, 1990)
- have a mechanism for renewal within itself
- possess an openness to the outside world so that it may respond to what is occurring there (Mills and Friesen, 1992)
- use systematic problem solving (Senge, 1991; Garvin, 1994)
- promote and continually experiment (Easterby-Smith, 1990; McGill and Slocum, 1993; Garvin, 1994)
- learn from their own experience and past history
- learn from the experience and best practices of others
- transfer knowledge quickly and efficiently throughout the organization (Garvin, 1994)
- possess accurate information systems
- have reward systems that recognize and reinforce learning
- have human resource practices that select people for their ability to learn
- possess a leader's mandate for unlearning and learning (McGill and Slocum, 1993) and
- effectively manage and use learning opportunities.

Hyland and Matlay (1997) assert that a learning organization can be defined or measured in terms of the sum total of accumulated individual and collective learning. However, Wang and Ahmed (2003) contend that organizational learning is not simply a collectivity of individual learning processes, but encompasses interaction between individuals within an organization, between organizations and between the organization and its contexts.

MEASUREMENT OF LEARNING ORGANIZATIONS

The following learning organization diagnostic instruments have been developed – the attributes of a learning organization (Goh, 2001), an instrument of a learning organization (Jashapara, 2003) and a market-based organizational learning measurement model (Morgan and Turnell, 2003). Additionally, the learning diagnostic questionnaire (LDQ) (Honey and Mumford, 1989), the learning climate questionnaire (LCQ) (Pedler *et al*, 1991), the learning environment questionnaire (Armstrong and Foley, 2003) and Hult *et al*'s (2002)

17-item learning climate measure have been developed to evaluate the learning climate of organizations.

IMPROVING EXPERIENTIAL LEARNING AT WORK

Learning organizations benefit from mechanisms that transfer learning from an individual to a group. Further, they must educate employees about how to learn and reward them for success in learning (Mills and Friesen, 1992). Work-based learning opportunities are crucially dependent on the way in which work is organized and allocated (Eraut, 1994). While Freedman (1967) states that learners are more influenced by their peers than by any other factor within their learning environment. For example, those factors that hindered learning within organizations were relations with other people, other people's characteristics, organizational structures, the environment and job characteristics (Vandenput, 1973).

Ideally, it is said that an organizational culture climate approach is needed, in which an organization encourages learning by encouraging individuals to identify their own learning needs and setting challenging learning goals; encouraging individuals to experiment; providing opportunities for learning both on and off the job; giving on-the-spot feedback; allowing time for employees to review, conclude and plan learning activities; and tolerating some mistakes, provided they try to learn from them (Mumford, 1986).

An organization's learning climate is associated with an employee's perception of the environment that is created by their organization's practices, procedures and reward systems that support learning (Schneider and Reichers, 1993; Cunningham and Iles, 2002). Learning environments should be both supportive and challenging (Knox, 1986), affording trust, mutual support, acceptance of the individual, warmth and respect, thereby enabling the learner to take risks, admit to difficulties and problems, give and receive feedback and cope with the allied stresses (Boydell, 1976). The 'ideal' working situation: 'has jobs that grow and expand; takes action to meet development needs; allows people to decide how to meet their objectives; has collaborative processes for setting objectives; diagnoses the causes of problems; encourages people to be open about their problems; welcomes new ideas; constantly changes; has top management who are actively involved in training and development activities; provides opportunities to use new skills; constantly strives to improve quality; encourages people to aim high; uses task forces and project teams; encourages people to experiment with new ways of doing things; actively supports people's plans to implement something learned on a course' (Honey and Mumford, 1989).

EMPIRICAL FINDINGS

SAMPLE

The sample investigated comprised students from the BSc in CM&QS programme at the University of Manchester/UMIST who had completed at least 46 weeks of SWE with a UK construction company (the majority of whom were sponsored by a consortium member). The sample comprised the years 1996–2003. Table 1 indicates the total number of students in each year and the number of responses for each year. Ultimately, 165 students out of a possible 185 took part, representing a response rate of 85%.

QUESTIONNAIRE

The students were required to complete a learning climate questionnaire (LCQ), which is an inventory designed to elicit information on whether they considered their work placement organization provided an appropriate climate. The LCQ required the students to rate 15 pairs of statements on a five-point semantic differential scale. The chosen statements were derived from Pedler *et al*'s (1991) measuring the quality of your learning climate, Honey and Mumford's (1989) work situation items and Mumford's (1980) ways in which

TABLE 1 **Number of responses, total number of students, % response and number of female students (per year and in total)**

	1996	1997	1998	1999	2000	2001	2002	2003	TOTAL
No. of responses	18	21	20	20	19	29	22	16	165
Total in cohort	22	22	24	24	23	36	26	17	194
% response	82	95	83	83	83	81	85	94	85
No. of females	2	5	5	4	8	9	7	2	42

supervisors can improve the learning climate. Also, the students were asked to list five positive and five negative aspects of their year in industry. The questionnaires were all completed in week 1, semester 1 of the student's final year of study.

ANALYSIS

Data analyses were undertaken using the Statistical Package for the Social Sciences (SPSS for Windows, release 10.0.7). Descriptive statistics were calculated for each item of the LCQ. The items were ranked based on the mean score. A three ('k') factor analysis was performed for the LCQ and factor scores generated. Each item of the LCQ and the LCQ summary variable was analysed for differences between the total sample and the median score (2) using t-tests and its comparable non-parametric test. They were further analysed together with the three factor scores for differences between sub-groups based on the year of study (1996–2003) by means of one-way analysis of variance (ANOVA) and its comparable non-parametric test. Further, each item of the LCQ, the LCQ summary variable and the three factor scores were correlated with time (year of study). Finally, the positive and negative factors that influenced the students' SWE were collated and ranked in order of importance.

INTERVIEWS

On completion of the analysis, unstructured interviews were conducted with six representatives (training/human resource managers) from five of the consortium of construction companies involved in the programme. These individuals were responsible for placing SWE students within their organization and monitoring the students' progress during the SWE.

PERCEPTIONS OF THE LEARNING CLIMATE

LCQ SUMMARY VARIABLES

Initially, principal components extraction with varimax rotation was used to determine the underlying dimensions of the 15 items of the LCQ. The number of factors extracted was dictated by Kaiser's criterion. This produced a three-factor solution, while a scree plot indicated that the true number of factors lay between two and four factors. Two-, three- and four-factor solutions were carried out, and after inspecting the factor loadings matrices, the three-factor solution was computed. The initial eigenvalues ranged from 5.71 for factor one to 1.13 for factor three and the solution accounted for 55.36% of the variance. The final solution was generated using principal factor extraction with an oblique (Oblimin) rotation. The three-factor solution accounts for 45.1% of the total variance in the LCQ. Variables were ordered and grouped by size and interpretative labels suggested.

Factor one – human support (HS) – is associated with the items 8, 10, 4, 11, 5, 7, 6 and 9: 'People are very willing and supportive; pleasure is taken in the success of others'; 'The organization is an open and friendly place'; 'People are usually ready to give their views and pass on information'; 'Discussion of problems is actively encouraged'; 'People are recognized for good work and rewarded for effort and learning'; 'If people develop a new skill or technique there is plenty of opportunity to use it'; 'People manage themselves and their work; there is great emphasis on taking personal responsibility'; 'Constructive feedback is often provided about your performance'.

Factor two – staff development systems (SDS) – is associated with the items 3, 1 and 2: 'There is a systematic process for identifying individual development needs'; 'There are lots of resources; development facilities are very good'; 'People are encouraged to learn at all times and to extend themselves and their knowledge'.

Factor three – working practices (WP) – is associated with the items 14, 15, 13 and 12: 'Accepts that some forecasts will prove to be inadequate'; 'Explicitly deals with risk and uncertainty'; 'Working practices and structures are constantly under review'; 'High standards are a goal to be achieved'. Three-factor scores were generated using the regression method. The validity of these dimensions is supported by Vandenput (1973), as discussed earlier.

Additionally, a weighted average LCQ summary variable was created.

DESCRIPTIVE STATISTICS

The alpha reliability estimate for the total scale was 0.86, while the split-half reliability estimate was 0.90. This suggests the inventory is internally consistent. Frequencies and summary statistics for the 15 statements used in the LCQ are presented in Table 2,

TABLE 2 Frequencies, means and standard deviations of individual items of the learning climate questionnaire (LCQ) (n = 165)

FIVE-POINT SEMANTIC DIFFERENTIAL SCALE	4	3	2	1	0		MEDIAN	MEAN	SD
People manage themselves and their work; there is great emphasis on taking personal responsibility	37	**79**	42	5	2	People conform to rules and standards at all times – no personal responsibility is taken or given	3	2.87	0.83
People are usually ready to give their views and pass on information	40	**77**	30	11	7	People tend to keep their feelings to themselves; are secretive and information is hoarded	3	2.80	1.02
The organization is an open and friendly place	49	**61**	35	9	11	There is little openness and support; the organization is cold and insular	3	2.78	1.13
Discussion of problems is actively encouraged	27	**59**	56	20	3	'People don't have problems'	3	2.53	0.97
People are encouraged to learn at all times and to extend themselves and their knowledge	26	**67**	39	25	8	There is little encouragement to learn; there are low expectations of people in terms of new skills and abilities	3	2.47	1.08
People are very willing and supportive; pleasure is taken in the success of others	24	**63**	51	19	8	People don't support each other; there is an unwillingness to pool or share information	3	2.46	1.03
High standards are a goal to be achieved	25	53	**59**	23	5	High standards are compulsory	2	2.42	1.01
Explicitly deals with risk and uncertainty	16	50	**74**	16	9	Avoids risk and uncertainty	2	2.29	0.96
There are lots of resources; development facilities are very good	23	**52**	49	30	11	Training packages, resources and equipment are limited	2	2.28	1.12
Working practices and structures are constantly under review	29	40	**48**	36	12	Working practices and structures are static	2	2.23	1.19
Accepts that some forecasts will prove to be inadequate	5	43	**90**	20	6	Does not accept inadequate forecasts	2	2.13	0.80
Constructive feedback is often provided about your performance	22	41	45	**46**	11	Constructive feedback is rarely provided about your performance	2	2.10	1.15
People are recognized for good work and rewarded for effort and learning	10	49	**64**	28	14	People's successes are ignored but blame is readily attributed	2	2.08	1.02
If people develop a new skill or technique there is plenty of opportunity to use it	8	50	**65**	30	12	If people develop a new skill or technique there are few opportunities to use it	2	2.07	0.99
There is a systematic process for identifying individual development needs	13	37	**51**	41	23	The identification of development needs is left to the individual	2	1.85	1.15

Note: Bold figure is the mode.

ranked based on their mean scores. Further, the items were tested for differences against the median value (2) (Table 3).

Those statements given a high rating: 'People manage themselves and their work; there is great emphasis on taking personal responsibility' (HS); 'People are usually ready to give their views and pass on information' (HS); 'The organization is an open and friendly place' (HS); 'Discussion of problems is actively encouraged' (HS); 'People are very willing and supportive; pleasure is taken in the success of others' (HS); 'People are encouraged to learn at all times and to extend themselves and their knowledge' (SDS); 'High standards are a goal to be achieved' (WP); and 'Explicitly deals with risk and uncertainty' (WP). The t-test for independent samples also revealed a very highly significant difference at the 0.1% level in these items when tested against the median value.

The results indicate that the working environment within construction organizations is perceived by the students to be supportive in terms of human support and to a lesser extent in working practices. However, the human support items appear to be related to the personal and informal support from colleagues. This finding is important as learning within an environment requires a human communications

TABLE 3 ANOVA and tests for differences in the learning climate questionnaire items (n = 165)

		MEDIAN (2)		YEAR OF STUDY	
	ITEM	T	Z	F	χ^2
Personal responsibility	HS	13.427***	–8.959***	0.719	4.782
People – information	HS	10.081***	–7.589***	1.135	8.832
Organization	HS	8.791***	–6.799***	2.011	12.630
Problems	HS	7.009***	–6.096***	2.101*	15.690*
Support	HS	5.728***	–5.062***	2.300*	15.687*
Encouragement to learn	SDS	5.625***	–5.038***	2.370*	18.690**
High standards	WP	5.412***	–4.949***	1.616	10.116
Risk and uncertainty	WP	3.881***	–3.527***	1.529	11.845
Resources	SDS	3.202**	–3.048**	1.922	13.777
Working practices	WP	2.491*	–2.554*	2.768**	19.000**
Forecasts	WP	2.050*	–1.951	0.698	4.036
Feedback	HS	1.150	–1.275	4.144***	25.000***
Recognition of work	HS	0.988	–0.829	1.764	12.392
New skills	HS	0.948	–0.825	1.278	9.095
Identification of needs	SDS	–1.619	–1.661	2.971**	19.677**
LCQ summary variable	LCQ	10.216***	–8.268***	3.290**	22.265**
Human support	FS	–	–	1.958	16.046*
Staff development systems	FS	–	–	3.681***	23.017**
Working practices	FS	–	–	1.406	10.448

***= p≤(0.001 **= p≤(0.01 *= p≤(0.05

T = t-test – one-sample test

Z = Wilcoxon signed ranks test

F = F Ratio one-way analysis of variance

χ^2 = chi-square Kurskal-Wallis 1-way ANOVA

HS = human support

SDS = staff development systems

WP = working practices

[Q35]

network or society (Rogers, 1986) and relates to the social context within which learning takes place (Lovell, 1980). Snell (1992) considers the main source of 'pain' in learning to be the prevailing organizational ethos of competitive individualism. Furthermore, Freedman (1967) states that learners are more influenced by their peers than by any other factor within their learning environment.

Those items that were not significantly different from the median value were: 'Constructive feedback is often provided about your performance' (HS); 'People are recognized for good work and rewarded for effort and learning' (HS); 'If people develop a new skill or technique there is plenty of opportunity to use it' (HS); and 'There is a systematic process for identifying individual development needs' (SDS).

These results suggested that the working environment was considered to be less supportive in terms of the more formal support given by managers within the organization and specifically in terms of the use of appraisal systems to identify development needs. This supports the findings of Scott and Harris (1998) who discovered that the majority of in-place project feedback systems were informal and unstructured, which prohibited effective learning from taking place.

DIFFERENCES IN THE STUDENTS' PERCEPTIONS OF THE LEARNING CLIMATE OF THEIR ORGANIZATION BASED ON THE YEAR OF STUDY

The items of the LCQ and its summary variable were tested for differences between sub-groups based on the students' year of study (see Table 1 for sizes of sub-groups). The results are presented in Table 3.

ANOVA revealed a very highly significant difference at the 0.1% level in the item – 'Constructive feedback is often provided about your performance' (HS) and the SDS factor score; a highly significant difference at the 1% level in the items – 'There is a systematic process for identifying individual development needs' (SDS), and 'Working practices and structures are constantly under review' (WP) and the LCQ summary variable; and a significant difference at the 5% level in the items – 'People are encouraged to learn at all times and to extend themselves and their knowledge' (SDS), 'People are very willing and supportive; pleasure is taken in the success of others' (HS) and 'Discussion of problems is actively encouraged' (HS).

Closer examination using Bonferroni's multiple comparison test revealed that for the following items:

- 'People are very willing and supportive; pleasure is taken in the success of others': there were no significant differences between sub-groups at the 5% level.
- 'There is a systematic process for identifying individual development needs': the mean score for the 1996 student group was significantly lower than that of the 2003 group (at the 5% level).
- 'Discussion of problems is actively encouraged' and 'People are encouraged to learn at all times and to extend themselves and their knowledge': the mean score for the 1996 student group was significantly lower than that of the 1998 group (both at the 5% level).
- 'Working practices and structures are constantly under review': the mean score for the 1996 student group was significantly lower than those of the 1997, 1998 and 2002 groups (all at the 5% level).
- 'Constructive feedback is often provided about your performance': the score for the 2001 student group was significantly higher than those of the 1996 and 1999 groups (both at the 1% level).
- The LCQ summary variable: the mean score for the 1996 student group was significantly lower than those of the 1997, 1998 and 2003 groups (all at the 5% level).
- The SDS factor score: the mean score for the 1996 student group was significantly lower than those of the 1998 and the 2002 groups (both at the 5% level) and the 2001 and 2003 groups (both at the 1% level).

The findings indicate homogeneity in the students' responses to 10 out of the 15 LCQ items. Furthermore, those items identified by ANOVA as being significant, with the exception of 'Constructive feedback is often provided about your performance', appear to reflect the dissatisfaction of the 1996 group. However, the results for this item may reflect a trend indicating an improvement over time.

RELATIONSHIPS BETWEEN THE LEARNING CLIMATE AND TIME

To investigate this further, the items of the LCQ questionnaire were correlated with time (1996–2003). Pearson's s (P's') and Spearman's rho (S'r') correlation coefficients were calculated and indicate that the following correlate significantly and positively with time:

- 'Constructive feedback is often provided about your performance' (HS): at the 0.1% level (P's' = 0.260; S'r' = 0.253).
- The SDS factor score: at the 1% level (P's' = 0.245; S'r' = 0.240).
- 'There are lots of resources; development facilities are very good' (SDS): at the 1% level (P's' = 0.235; S'r' = 0.232).
- 'There is a systematic process for identifying individual development needs' (SDS): at the 1% level (P's' = 0.213; S'r' = 0.220).
- The LCQ variable: at the 5% level (P's' = 0.171; S'r' = 0.175).
- 'The organization is an open and friendly place' (HS): at the 5% level (P's' = 0.166).

To remove the effect of the 1996 student group Pearson's s (P's') and Spearman's rho (S'r') correlation coefficients were calculated for the years 1997–2003 and revealed that only the following item correlated significantly and positively with time:

- 'Constructive feedback is often provided about your performance' (HS): at the 5% level (P's' = 0.178; S'r' = 0.176).

While the following item correlated significantly and negatively with time:

- 'Discussion of problems is actively encouraged' (HS): at the 5% level (S'r' = 0.174).

The results indicate that the students' scores for the item related to the provision of feedback has increased significantly over time. However, for the items relating to the provision of resources, support and identifying development needs, the SDS factor score and the average of all the LCQ variables, again, the results would appear to indicate that the 1996 group were particularly dissatisfied with these items rather than demonstrating an improvement trend over time.

POSITIVE AND NEGATIVE SWE FACTORS

Table 4 presents positive and negative factors that influenced the students' SWE ranked in order of importance for two representative student groups – 1998 and 2002. The responses of the two student groups are very similar. For example, level of responsibility, experience and variety of work and a friendly atmosphere were consistent positive factors.

TABLE 4 **Factors influencing the students' perception of supervised work experience**

POSITIVE FACTORS			
1998		2002	
1	Level of responsibility	1	Level of responsibility
2	Experience of work	2	Variety of work
3	Friendly atmosphere	3	Experience of work
4	Working relationships	4	Friendly atmosphere
5	Training opportunities	5	Good/regular salary
6	Variety of work	6	Working in teams
7	Social life	7	Link between studies and work
8	Involvement in prestigious project	8	People pass on information
9	Dealing with people	9	Training opportunities
		10	Working relationships
NEGATIVE FACTORS			
1998		2002	
1	Lots of travel/being away from home	1	Long hours
2	Poor site management	2	Boring/less glamorous tasks
3	Lack of money	3	Lots of travel
4	Long hours	4	Poor supervision
5	Poor training	5	Poor training
6	Poor supervision	6	Lack of cooperation/trust
7	Lack of variety of work	7	Weather conditions/working on site
8	Lack of understanding of your ability	8	Sexism
9	Supervisors ill prepared	9	Being the student
10	Poor working conditions	10	Uncertainty/lack of choice of job location

Likewise, lots of travel, long hours, poor training and poor supervision were consistent negative factors. It is interesting to note that for the first time in 2002, the students perceived the salary they received to be a positive influence rather than a negative factor as indicated in 1998. Disturbing, however, is that students considered their tasks to be boring or less glamorous and that several comments were made regarding sexism encountered within construction organizations.

INTERVIEWS

All the interviewees viewed sponsorship and SWE in particular as a key component of graduate recruitment – 'an opportunity to build a relationship with a prospective employee' – with conversion rates of between 60% and 96% being reported for 2003. In an ideal world, most reported that they would not recruit a student who had not undertaken SWE.

Generally, SWE posts are driven by a business need with the placement paid through a contract, rather than a training budget, enabling the students to be integrated into a business area. However, for one organization the SWE was funded equally through a project and the training budget enabling the SWE student to be controlled by the training department, thereby, allowing them greater control over what the SWE student did.

Within all the organizations, SWE students were inducted on to structured training schemes during the SWE and received formal training, for example, in site safety and site supervision. Furthermore, two organizations operated rotational placement schemes with students working in two or three different areas of the organization. Others explained that they had embraced NVQ level 3, which is linked to a framework for the SWE with the intention that the student obtains a Construction Skills Certification (CSCS) card. However, one representative stated that his organization had rejected NVQ as not being readily applicable to the construction industry, while another suggested several incentives for completing the NVQ, for example, it counted in lieu of a diary required by professional institutions and completion of the NVQ during SWE meant that when the student returned to the sponsoring company on graduation they were deemed to have completed year one of the organization's graduate scheme, therefore, they could be employed at a higher level.

Commonly, SWE students are allocated a learning adviser (a training or human resource function) and also a supervisor (line manager/mentor/coach). Jointly, the learning adviser and supervisor are responsible for ensuring appropriate work experience. The learning adviser usually visits the students between two and five times during the SWE to monitor progress; this is in addition to various line management appraisal schemes. Culturally, many senior managers within the organizations had undertaken year-out placements themselves; therefore, the majority are willing to supervise students. However, most interviewees suggested that they selected specific managers and projects to allocate students to. Additionally, two organizations provided supervisors with people skills training.

Whereas most organizations used formal questionnaires to obtain information on how students had performed on the SWE, only one organization used a formal questionnaire to elicit the students' views on their work placement, two organizations employed exit interviews and the others relied on informal feedback from the student.

'No surprise' was the general consensus over the results of the learning climate questionnaire, although it was suggested that by adopting the NVQ scheme, the scores for working practices should improve. Likewise, personal development was held to be the responsibility of the individual, although one interviewee thought that the identification of the students' learning needs was covered by their appraisal scheme. Surprisingly, some organizations excluded SWE students from their appraisal scheme.

Similarly, there was general acceptance of the positive and negative aspects of SWE. For example, long hours were accepted as being industry-wide and a cultural issue, while 'boring tasks' were seen as a feature of all jobs, although one interviewee commented that this response was possibly linked to high student expectations. Finally, sexism was held to be unacceptable, although one interviewee suggested that it was necessary to distinguish between banter and harassment, with a need to set a balance between the comfort of the student and not alienating existing staff.

CONCLUSIONS

The following conclusions have been drawn from the investigation.

THE LEARNING CLIMATE WITHIN CONSTRUCTION ORGANIZATIONS (WORKING ENVIRONMENT)

- The working environment is perceived to be supportive in terms of human support and to a lesser extent working practices. These human support items appear to be related to the personal and informal support from colleagues.
- It is considered to be less supportive in terms of the more formal support given by managers within the organization and specifically in terms of the use of appraisal systems.
- There is a high degree of homogeneity in the responses of the students to the items of the LCQ questionnaire.
- There has been a significant improvement over time in the students' perception of the provision of feedback on individual performance within construction organizations, however this item is still ranked 12th out of 15 LCQ items.

Construction organizations should consider either introducing, or applying more effectively, formal appraisal systems, as a mechanism for identifying individual development needs. Likewise, they should critically examine how managers of SWE students provide formal support. For example, they should consider introducing effective feedback mechanisms that require both the individual to critically reflect on their own performance and the organization to provide effective constructive feedback on an individual's performance.

FACTORS INFLUENCING THE SUPERVISED WORK EXPERIENCE

- Level of responsibility, experience and variety of work, and a friendly atmosphere were consistently considered to be positive factors.
- Lots of travel, long hours, poor training and poor supervision were consistently considered to be negative factors.
- For the first time in 2002, the students perceived the salary they received to be a positive influence rather than a negative one as indicated in 1998.
- Current negative influences were: students considered their tasks to be boring or less glamorous, and the sexism encountered within construction organizations.

The implication of these findings is that construction companies need to accurately assess their ability to provide an effective learning environment and to address any deficiencies. This would appear crucial if they are to attract potential employees to a career in construction and to retain students within the industry after SWE.

AUTHOR CONTACT DETAILS

Dr David Lowe: Management of Projects Research Group, School of Mechanical, Aerospace and Civil Engineering, University of Manchester, PO Box 88, Manchester, M60 1QD, UK.
Tel: +44 (0) 161 306 4643, fax: +44 (0) 161 306 4646,
e-mail: david.lowe@manchester.ac.uk

REFERENCES

Armstrong, A. and Foley, P., 2003, 'Foundations for a learning organization: organization learning mechanisms', in *The Learning Organization*, 10(2), 74–82.

Ashworth, P. and Saxton, J., 1992, *Managing Work Experience*, London, Routledge.

Binsted, D., 1980, 'Design for learning in management training and development: view', in *Journal of European Industrial Training*, 4(8), 1–32.

Boydell, T., 1976, *Experiential Learning*, Manchester, Manchester Monographs.

Cavill, N., 1999, 'Where have all the young QSs gone?', in *Building*, 264(12) 24–25.

Cunningham, P. and Iles, P., 2002, 'Managing learning climates in a financial services organization', in *Journal of Management Development*, 21(6), 477–492.

Dainty, A.R.J. and Edwards, D.J., 2003, 'The UK building education recruitment crisis: a call for action', in *Construction Management and Economics*, 21(7), 767–775.

Davies, L., 1990, *Experienced-Based Learning within the Curriculum: A Synthesis Study*, Sheffield, Association of Sandwich Education and Training/CNAA.

Easterby-Smith, M., 1990, 'Creating a Learning Organization', in *Personnel Review*, 19(5), 24–28.

Eraut, M., 1994, *Developing Professional Knowledge and Competence*, London, The Falmer Press.

Freedman, M.B., 1967, *The Student and Campus Climates of Learning*, Washington, US Department of Health, Education and Welfare.

Gardiner, P., 1999, 'Soaring to new heights with learning oriented companies', in *Journal of Workplace Learning: Employee Counselling Today*, 11(7), 255–265.

Garratt, B., 1999, 'The learning organization 15 years on: some personal reflections', in *The Learning Organization*, 6(5), 202–207.

Garvin, D.A., 1994, 'Building a learning organization', in *Harvard Business Review*, 71(4), 78–91.

Goh, S.C., 2001, 'The learning organization: an empirical test of a normative perspective', in *International Journal of Organizational Theory and Behaviour*, 4(3/4), 329–355.

Hampton, J., 2001, 'Fishing for competence', in *Construction Manager*, September 18–19.

Harri-Augustein, S. and Thomas, L., 1991, *Learning Conversations: The Self-organized Learning Way to Personal and Organizational Growth*, London, Routledge.

Honey, P. and Mumford, A., 1989, *The Manual of Learning Opportunities*, Maidenhead, P. Honey and A. Mumford.

Hult, G.T.M., Ferrell, O.C. and Hurley, R.F., 2002, 'Global organizational learning effects on cycle time performance', in *Journal of Business Research*, 55(5), 377–387.

Hyland, T. and Matlay, H., 1997, 'Small businesses, training needs and VET provision', in *Journal of Education and Work*, 10(2), 129–139.

Jackson, G. and Wirt, J., 1996, 'Putting students to work', in *Training and Development*, 50(11), 58–60.

Jashapara, A., 2003, 'Cognition, culture and competition: an empirical test of the learning organization', in *The Learning Organization*, 10(1), 31–50.

Knox, A.B., 1986, *Helping Adults Learn*, San Francisco, Jossey-Bass.

Love, P., Li, H., Irani, Z. and Faniran, O., 2000, 'Total quality management and the learning organization: a dialogue for change in construction', in *Construction Management and Economics*, 18(3), 321–331.

Lovell, R.B., 1980, *Adult Learning*, London, Croom Helm.

Mann, W., 2002, 'Profile: Rick Willmott', in *Contract Journal*, 416(6400), 15–16.

Marsick, V.J. and Watkins, K.E., 1990, *Informal and Incidental Learning in the Workplace*, London, Routledge.

McGill, M.E. and Slocum, J.W., 1993, 'Unlearning the organization', in *Organizational Dynamics*, 22(2), 67–79.

Mills, D.Q. and Friesen, B., 1992, 'The Learning Organization', in *European Management Journal*, 10(2), 146–156.

Morgan, R.E. and Turnell, C.R., 2003, 'Market-based organizational learning and market performance gains', in *British Journal of Management*, 14(3), 255–274.

Mumford, A., 1980, *Making Experience Pay – Management Success through Effective Learning*, London, McGraw-Hill.

Mumford, A., 1986, 'Learning to learn for managers', in *Journal of European Industrial Training*, 10(2), 1–28.

Mumford, A., 1987, 'Learning styles and learning', in *Personnel Review*, 6(5), 20–23.

Mumford, A., 1992, 'Individual and organizational learning: the pursuit of change', in *Management Decision*, 30(6), 143–148.

Padaki, V., 2002, 'Making the organization learn: demystification and management action', in *Development in Practice*, 12(3/4), 321–337.

Pedler, M., Burgoyne, J. and Boydell, T., 1991, *The Learning Company: A Strategy for Sustainable Development*, London, McGraw-Hill Book Company.

Quality Assurance Agency, 2001, *Code of Practice: Placement Learning*, Gloucester, QAA.

Rogers, A., 1986, *Teaching Adults*, Buckingham, Open University Press.

Salaman, G., 2001, 'A response to Snell – the learning organization: fact or fiction?', in *Human Relations*, 54(3), 343–359.

Schneider, B. and Reichers, A.E., 1993, 'On the etiology of climates', in *Personnel Psychology*, 36(1), 19–39.

Scott, S. and Harris, R., 1998, 'A methodology for generating feedback in construction', in *The Learning Organization*, 5(3), 121–127.

Senge, P.M., 1991, 'Learning organizations', in *Executive Excellence*, 8(9), 7–8.

Snell, R., 1992, 'Experiential learning at work: why can't it be painless?', in *Personnel Review*, 21(4), 12–26.

Tsang, E.W.K., 1997, 'Organizational learning and the learning organization: a dichotomy between descriptive and prescriptive research', in *Human Relations*, 50(1), 73–89.

Vandenput, M.A.E., 1973, 'The transfer of training: some organizational variables', in *Journal of European Training*, 2(3), 251–262.

Wang, C.L. and Ahmed, P.K., 2003, 'Organizational learning: a critical review', in *The Learning Organization*, 10(1), 8–17.

Work Experience Group, 2002, *Work Related Learning Report*, Nottingham, DFES Publications.

ARTICLE

Investigating the Synergy between Teaching and Research in a Teaching-led University

The Case of an Architectural Technology Undergraduate Programme

Stephen Emmitt

Abstract

This paper seeks to investigate the theoretical and practical links between teaching and research in a teaching-led university in the UK. Focus is on the new architectural technology undergraduate programmes that, in theory at least, provide an opportunity to integrate research and teaching. The approach used was a case study, supported by a small amount of data collection from other institutions. An extensive literature review demonstrated the benefit to both students and academic staff of incorporating research into the curriculum. The data collection comprised semi-structured interviews with course leaders, managers and students, together with the monitoring of changes to an architectural technology programme over four years. The interviews were concerned with individuals' perceptions. The monitoring was concerned with the practical application of research to the curriculum. The data collection highlighted a series of issues, ranging from ineffective communication, poor resourcing and inadequate practical support for research. Research integration appeared to be wholly dependent on, and driven by, academics' personal search for synergy. Although the research reported here may, on first sight, appear a little negative, it would appear that with clearer direction from the university's senior managers, together with a re-evaluation of priorities at line-manager level, many opportunities could be realized without additional resourcing. From the issues identified, a number of areas for improvement are highlighted for future consideration.

■ *Keywords* – Architectural technology; ethnography; integration; research; teaching

INTRODUCTION

There are two views widely held by staff in teaching-led universities about teaching and research. Namely, that research is important and should be closely linked with teaching, or conversely that teaching is paramount and research is an unwelcome distraction. The Research Assessment Exercise (RAE) puts pressure on universities to encourage lecturers to become researchers, thus helping to secure funding via research output. This has placed pressure on staff time and has led to competent lecturers feeling undervalued if they do not have publications to their name. It has also led to some rather polarized views about research and its relationship to student learning. These pressures are particularly acute in teaching-led universities (Smith and Brown, 1995). Barnett (1992) highlighted the *ad hoc* nature of the connection between teaching and research in the polytechnics (now teaching-led universities). The manner in which an academic's research activities find their way into the curriculum was seen as important; were they simply 'bolted on' or were they integral to the work of the students? Architectural technology undergraduate degree programmes in the UK are being offered in educational establishments that have recently acquired university status, or are actively pursuing university status. The first degree in architectural technology was launched in the early 1990s, since which time the number of programmes has grown to 27 (BIAT, 2004). This new undergraduate programme provides a useful vehicle with which to investigate the link between teaching and research in built environment education. The scope of the research reported here was to observe, describe and critically evaluate how research was integrated into the curriculum of an architectural technology undergraduate programme.

ARCHITECTURAL TECHNOLOGY – A 'NEW' DISCIPLINE

The development of the architectural technology discipline can be traced back to the Oxford conference of 1958, which proposed the abolition of pupilage and part-time courses for architects and with it the formal creation of the architectural technician discipline. This essentially created a two-tier system (Crinson and Lubbock, 1994) – those responsible for controlling design (architects) and those with practical skills (the architectural technicians).

The Royal Institute of British Architects' (RIBA) 1962 report *The Architect and His Office* identified the need for an institution (other than the RIBA) that technicians could join to ensure maintenance of standards for education and training (RIBA, 1962). Technical design skills were identified as a missing component of architectural practice and the report urged the diversification of architectural education, suggesting that architects who chose to specialize in technology (rather than design) – the 'architechnologists' – should still be allowed to join the RIBA (RIBA, 1962). The report acknowledged that technicians were needed in architects' offices to raise productivity and standards of service, for which they would require education and training in the preparation of production information and technical administration; 'design' was specifically excluded from the technicians' training.

In 1965, the Society of Architectural and Associated Technicians (SAAT) was formed and inaugurated as an Associated Society of the RIBA under Byelaw 75 of the RIBA's charter in 1969 (SAAT, 1984). In May 1984, the SAAT published an influential report, *Architectural Technology: The Constructive Link,* which drew on the RIBA's earlier report to develop a view of construction for the 1980s and beyond. The report was important in establishing a sense of identity for architectural technicians, as a role complementary to architects. In particular the 'constructive link' has since been seen as an important concept, at the heart of the modern architectural technology discipline, implicit in the government-led reports by Latham (1994) and Egan (1998).

From inception in 1965, SAAT quickly became established, changing its name to the British Institute of Architectural Technicians and then to the British Institute of Architectural Technologists (BIAT). With the change of name from technician to technologist came the promotion of degree-level qualifications for its members. In 2005, the institute received a royal charter and changed its name to the Chartered Institute of Architectural Technologists (CIAT). Over a quarter of CIAT's 6500-plus members practise privately or are running businesses either as partners or co-directors with architects and other building professionals, often providing services in direct competition with architectural and surveying practices. The trend is similar to that in the US where there has been, and continues to be, a shift from the architect to the architectural technologist in terms of responsibility for building detailing and management.

BIAT has developed a common framework in conjunction with the RIBA that allows students to transfer relatively easily (theoretically at least) from architecture to architectural technology or vice versa. One institution has developed a common first year for architects and technologists, with students making an informed choice after their first year of study, taking a design- or technology-rich pathway for their final two years. Universities that employ modular schemes have found it more difficult to meet the common framework, since their architectural technology degrees have more synergy with their building surveying and construction management programmes. In May 2000, the QAA benchmarking document for architectural technology, developed in conjunction with BIAT, was published (QAA, 2000). Neither the QAA architectural technology benchmarking documentation (QAA, 2000) nor BIAT's (BIAT, 2000) guidelines for submitting a degree programme for validation make any specific reference to research. The validation document requests a list of staff associated with the degree and for a list of their research interests and a list of their publications from 1992 onwards. CIAT's Innovation and Research Committee was set up (1996) in an attempt to bring together disparate research under the common umbrella of architectural technology and promote research to the membership.

There is a view within the professional accrediting institution that research has a place in the undergraduate programmes, both incorporation of research findings and the teaching of research methods via the dissertation (Brown, 1996, 2000; Emmitt, 2000, 2001; Brookfield *et al*, 2004). However, the manner in which research output relates to specific undergraduate programmes is not particularly clear.

THE RELATIONSHIP BETWEEN TEACHING AND RESEARCH

In the UK, the 'university model' of higher education is founded on the principle that research and teaching are mutually dependent (e.g. Barnett, 1990) and thus they tend to reinforce one another, indeed the widely held view of higher education is that teaching is done by those who are research active (Ball, 1992). While this model may work for the established (research-led) universities, the new (teaching-led) universities have little tradition or ethos of research (Ball, 1992). Although the new universities are trying to address the issue (driven by funding pressures and the RAE) they still have a long way to go in trying to change their culture and catch up with those longer established. With increased emphasis on higher education providing for the majority, rather than the minority, there is an argument that programmes should link to the capabilities and wishes of a broad range of students in which research has no part (e.g. Ball, 1992). At the other end of the scale, research is seen to be important in promoting teaching quality (Barnett, 1992; Perkins, 1998). The perception that there is a link appears to be stronger than the actual link (Brew and Boud, 1995).

Literature that has sought to address the link, perceived or otherwise, between academic research and teaching tends to be heavily influenced by the background of the author(s), with the established universities arguing for the connection, the newcomers arguing against. With increased demand for higher education and the increase in costs of provision, there is considerable pressure to find a cheaper method of delivery and, because research is expensive to produce, it has come under increased pressure. Many managers would, according to Lindsay (1998), be very relieved if the literature categorically proved that teaching could get along fine without research.

In their overview of (mainly US) research into research productivity and student ratings of teaching, Pascarella and Terenzini (1991) found a very low positive relationship at an individual level. Ramsden and Moses (1992) found no evidence of a simple functional relationship between research output and the effectiveness of teaching. Astin's (1993) study of US colleges found students to be more dissatisfied in the research-orientated colleges than their peers in the teaching colleges. The greater the research output, the greater the dissatisfaction of students. These studies are potentially misleading because it is only a small minority of individuals within universities who produce the majority of the research output; the majority of staff are primarily involved in teaching (Ramsden, 1998).

Feldman's (1987) review of numerous studies found that there was a positive link, but that statistically it was insignificant. Qualitative studies by Neumann (1994) in Australia and Jenkins *et al* (1998) in England found a positive relationship between research-active staff and student satisfaction. Here, the student perception was that research-active staff were incorporating research into their teaching and this helped to make the courses current and intellectually exciting. The studies also found that incorporating current research was not a substitute for good teaching practice and that staff time needed careful management to minimize the negative effect that absent staff (research sabbaticals, conference attendance, etc.) may have on the student experience.

Lindsay (1998) reported similar findings. In studies examining the experiences of 250 students it was found that both undergraduates and postgraduates believed that research-active lecturers make more successful teachers (Utley, 2001). In an analysis of existing literature, Brew and Boud (1995) came to the conclusion that the arguments put forward about the link between teaching and research had not progressed very much in recent years and had become sterile. Instead, they argue that we should be looking at the relationship between learning and research, concluding that research and learning cannot be separated.

A PROBLEM WITH DEFINITIONS

In the work reported above, there is no overriding definition of the term 'research', thus it is particularly difficult to compare findings from the different studies. Ball (1989, 1992) suggests that unless a distinction is made between different kinds of research, the dilemma cannot be resolved. He uses three definitions. Scholarship: it is the duty of all who teach in higher education (regardless of type of institution) to engage in scholarship and hence enhance the quality of teaching. The second and third distinctions – fundamental research and contract research (both are undertaken outside of, as well as within, higher education institutions) – he argues do not necessarily enhance teaching, although they may do so, and as such neither is a 'necessary condition' of higher education. Ball's argument is that with increased access to higher education, we need to reconsider the 'fundamental link' between teaching and research because more students bring different pressures. Colbeck (1998) has shown that the perception of what constitutes research affects the possibility of staff seeing a relationship between their teaching and their research. She found that staff in the less well research-funded universities saw stronger possibilities for linkage than their colleagues in the better-funded institutions. This was accounted for by the fact that the academics in the better research-funded universities were concerned with high prestige research (which was difficult to link to teaching), while those less well funded saw research as the production of textbooks and teaching aids (which has clear links). Defining research is important in the context of this inquiry, but is fraught with difficulties. Where, for example, is the line between fundamental research and scholarly activity to be drawn? For built environment disciplines, the picture is more complex because the courses contain an element of design and there has been some debate as to whether or not design constitutes legitimate research activity (e.g. Yeomans, 1996). Such distractions do not help the argument here. So, for the purposes of this study, the definition used is that embedded in teaching-led university staff contracts, i.e. it is used in its widest sense to include scholarly activity.

RESEARCH AS PART OF THE STUDENT EXPERIENCE

Brew and Boud (1995) put forward a convincing argument for the correlation of teaching and research, suggesting it could be achieved through 'exploitation' of the link between research and teaching in the design of course curricula. Garnett and Holmes (1995) develop a similar argument based on the view that within each institution there should be a balanced relationship between teaching and research, i.e. teaching and research should complement one another. They put forward a convincing thesis for students to be working in an environment that is 'impregnated' with a research culture, but note that academics are finding it increasingly difficult to find the time to maintain viable research activity. A more pragmatic view can be found in a paper by Race (1995) who notes the tension between academic research and student-centred teaching and learning. For him, the issue is about academics who are intent on developing the quality of student learning, arguing that the most effective teachers are researchers. Earlier work found that the application of research to teaching could transform the teaching and learning experience (Grew, 1992), bringing benefits to both lecturer and student.

Perkins (1998) makes the point that staff rarely articulate the link, yet there is an underlying philosophy that underpins the commitment to integrate research and teaching. This argument has gained greater importance as we enter the knowledge economy (e.g. Jacob and Hellstrom, 2000) with employers looking for students who know how to apply knowledge (e.g. Allee, 1997). So the challenge is not so much in incorporating research into the curriculum, rather it is in encouraging students to engage in the creation and application of knowledge, i.e. they need to develop research skills (Barnett, 1990). Nord (1996) has described research and teaching as processes characterized by 'continuous learning'. Drawing on this, Frost (1997) suggests that as long as we approach teaching or research with an open mind and with an expectation that creativity, discovery and improving understanding are taking place, then we are continuing to learn. With this approach, it is then possible for the twin crafts of teaching and research to 'feed and enhance' each other. Reflecting on a career as an academic, Aldrich (1997) posits that both faculty staff and students benefit from having research findings and the research process explained to them in the classroom. He argues that it is through such exposure to research that they will become more critical judges of the information before them.

RESOURCE IMPLICATIONS

Staff need time to stay up-to-date with their field of study but, according to Barnett (1990), should not be obliged to undertake research. Fukami (1997) has discussed the challenge of trying to integrate research, teaching, administration and a social life. She claims that the easy and less stressful option is to concentrate on one thing at a time. The UK government has suggested that universities should be separated and funded on the basis of their research, thus forming 'research' universities and 'teaching' universities (Advisory Board for the Research Councils, 1987). Kogan and Henkel (1992) have discussed the constraints brought to bear on research-active staff – the main ones being time, academic plans, performance indicators and cost centres. Their point is that this is not a simple matter and care is required when considering resources. This helps to underline the role of line managers in allocating resources and the importance of staff development as demonstrated by Harris (1995).

METHODOLOGY

The literature review helped to highlight the complexities inherent in trying to investigate the link between teaching and research. The implication for the data collection exercise is that a variety of stakeholders, i.e. course leaders, managers and students, will have a view and this must be addressed. The literature review established a clear preference for quantitative methods – an approach questioned by some of the more recent research. A methodology was required that allowed appropriate data collection to be achieved within a limited time frame, while at the same time working within an ethical framework. The specific aims of the research were to gather the opinions of individuals about their approach to research and teaching and try to observe how research is incorporated into the curriculum, which lends itself to qualitative research techniques (e.g. Rosen, 1991; Cohen and Manion, 1994; Gill and Johnson, 1997; Symon and Cassell, 1998; Seale, 1999).

Access to data was an important consideration. At the time, the author was employed as a course leader of an architectural technology programme at a teaching-led university in the UK, thus allowing the opportunity to engage in participant observation. Since the intention was to interview and monitor the author's peers and students, awareness and attention to ethical issues was paramount. Complete transparency was required throughout all stages of the research. So too was permission from staff, students and the management of the university, which was sought and was forthcoming.

In qualitative research, it is recognized that the distinction between data collection and analysis may not necessarily be clear cut because the interviewer will be testing and analysing the data as interviews progress, i.e. it is an iterative process (Potter, 1996). In recognition of this, the data collection was designed to include an iterative element. A multi-method approach (which constitutes a form of triangulation) was adopted, centred on the monitoring of a specific module and course at a new university in the UK, supported by a review of documentation, semi-structured interviews and questionnaire surveys over a period of four years.

THE INTERVIEWS: OPINIONS AND PERCEPTIONS

Opinions were gathered through semi-structured, face-to-face interviews with course leaders, line managers and students using an iterative approach. Eight course leaders were interviewed, four at the case study institution who were responsible for directing built environment courses within the modular scheme and four architectural technology course leaders at other higher education institutions, which also operated under modular schemes. This allowed for comparison of views and hence a modest check on the validity of the views expressed at the case study institution. All had a minimum of five years' experience as course leader, four claimed to be research active and four claimed to be 'too busy' to engage in research.

The responses were analysed and the pertinent issues incorporated into the questions for the semi-structured interviews with the managers and the students. Four managers at the case study institution were then interviewed based on the comments from all of the course leaders, thus helping to overcome ethical concerns and going some way to protect anonymity. Eight final-year (level 3) students were then interviewed, four from the architectural technology programme and one from each of the four other built environment programmes. Combined, the interviews provided a useful indication of the opinions and perceptions of a wide range of people associated with the architectural technology undergraduate programmes, briefly summarized below.

THE COURSE LEADERS' VIEWS

The views of the course leaders at the case study institution did not differ from those interviewed at the other institutions. Views were consistent throughout, the only differences of opinion related to whether or not the individual was research active, and even here the differences were relatively subtle. This suggests that the views held at the case study institution were not specific to that institution only.

Management and organizational structure

All course leaders felt that a balanced team of researchers and teachers was needed, but then went on to explain why such an approach was not working very well. Managers were criticized for not giving direction or the opportunity to discuss integration. One reason given for this was that the majority of line managers were not research active and so they were less sympathetic to research, summed up by the comment: 'There is a short-sighted view at managerial level with regard to research and its importance on undergraduate degrees.' Research units were seen as 'peripheral' to teaching and the research output difficult to link into teaching. Communication between those charged with managing teaching and those managing research was perceived to be very poor.

Curriculum and research-based learning

The consensus was that the majority of staff research activity had a tenuous relationship to the undergraduate programmes. The importance of research and its incorporation into teaching was recognized, but those interviewed felt that integration was only being achieved in a small number of modules and more should be done to incorporate the department's collective knowledge. The drive to incorporate research was driven by individuals through their lectures. Interviewees claimed that students appreciated those lecturers who made the effort to incorporate research. Six of the eight course leaders claimed that they tried to incorporate research into their courses; the other two claimed a lack of time prevented them from doing so. The influence of individuals to 'push' for integration was noted and the cultural problems inherent in the teaching-led universities was also raised as an inhibiting factor.

Inclusive culture

The perception was that research had had relatively low importance in the past, but that more emphasis would be given to research in the future because of demands from employers and professional bodies. Seven of the course leaders wanted to see greater integration of research and teaching, one claimed to be happy with the present balance. The consensus was that the university departments did not make enough effort to integrate research knowledge with teaching activities. Two course leaders said that they would welcome clearer and more specific guidance from the accrediting bodies (e.g. BIAT, Chartered Institute of Building). One interviewee claimed that it was pointless talking about research integration until there was 'more sensitive and realistic deployment of duties', noting that, each year, staff were expected to do more work with less resources, thus research had to be undertaken outside formally scheduled duties, i.e. in academics' own time. Lack of time and lack of support from line managers was perceived to be a major barrier to achieving an inclusive culture, summed up in the comment: 'Our line managers are only interested in teaching, they think research should be done in our own time – there is a cultural and managerial problem here.'

THE LINE MANAGERS' VIEWS

Interviews were conducted with four managers. Their roles were head of department, manager of the undergraduate modular scheme, manager of the department's research activity and the group manager responsible for the architectural technology programme. Each contributed to policy issues with regard to research, teaching and/or staff deployment. Comments given were consistent between individuals, despite their different roles. All four taught on the modular programme, only the research manager claimed to be research active.

Management and organizational structure

Managers' reluctance to allocate adequate time to research was put to the managers. They responded by agreeing with the course leaders, justifying their response from a financial perspective. They held the view that the university did not have a strong research culture and did not generate very much money via research activity, therefore senior managers were reluctant, or simply unable, to allocate resources to research. The managers were critical of the university's senior management, looking to them for clear direction, and claiming that guidance was not forthcoming. It was acknowledged that research was low on the managers' list of priorities – teaching and administration took priority. All four managers blamed their senior managers (deans) for the problem, none of them saw it as a problem they could address unless directed to do so.

Curriculum and research-based learning

The managers felt that the department's research activities did relate to individual courses, both at undergraduate and postgraduate level. They recognized that individuals were responsible for driving this, mainly through their module delivery, and that there was no departmental policy on research and teaching integration. There was also recognition that some of the research carried out needed to be better related to the courses. The view was that professional bodies were starting to exert an influence, albeit a small one, with regard to incorporating research into undergraduate programmes, but that they should give clearer guidance.

Inclusive culture

The course leaders had claimed a lack of integration (lack of interest) between teaching and research at managerial level. Justification given for the lack of attention was inadequate financial support, although it was acknowledged that the university had encouraged research through research sabbaticals, staff development and internal promotions based on research output. The managers were quick to point out that the university was a 'teaching institution' and therefore academics should not expect much support for research. They had to deliver a teaching portfolio within a limited budget and with diminishing resources. Lack of time was viewed as the main factor that prevented adequate communication and knowledge sharing.

THE STUDENTS' VIEWS

Interviews were conducted with eight final-year students. At the start of the interviews, care was taken to explain how the term research was used in the context of the interviews. In contrast to the other interviews, these interviews were all short in duration because the students had little to say about research and teaching (thus prompting a second attempt at collecting data, see the case study).

Management and organizational structure

There was a clear plea from all students for improved communication between the academic staff and students; they wanted to know what research individual academics were doing. Students were very positive about the architectural technology programme and the majority of staff who taught on it.

Curriculum and research-based learning

The students were concerned with the applicability of their degree to their chosen profession and were aware of the benefits that research incorporation could give them, but they were critical of some staff for not doing more to keep them informed. When asked to name their favourite modules, the students were quick to identify individuals as being up-to-date and knowledgeable about their subject area. All of the lecturers named were research-active members of staff, which supports earlier findings (e.g. Jenkins *et al* 1998). When asked whether research skills were perceived to be important when entering the industry, there was a general consensus that they were. They saw exposure to research and the development of research skills as important factors for bringing about change in construction. Research skills were also identified as important for improving innovation, sustainable building and vital for areas such as product selection.

Inclusive culture

All students said that there should be a mix of staff skills, a balance between research and teaching staff. The students claimed that those lecturers who were research active gave better lectures and tutorials than those perceived to be teaching directly from textbooks. Research was seen to be important in keeping the degree programme lively, relevant and bettering the student experience. These views are consistent with those recorded by others (e.g. Grew, 1992).

THE MONITORING PROCESS: APPLYING RESEARCH IN PRACTICE

To investigate what was being done in practice, it was necessary to look in detail at the curriculum of an architectural technology undergraduate programme and its delivery. The case study monitoring period ran for 48 months using participant observation, supported by analysis of course documentation and a further series of short (semi-structured) interviews with teaching staff to help explain some of the observations. Here, the intention was to try and determine the actual extent of research incorporation within individual modules compared with the comments gathered from the interviews.

THE ARCHITECTURAL TECHNOLOGY CURRICULUM

The undergraduate programme in architectural technology was launched in the mid-1990s and is part of a modular undergraduate scheme. Since its launch, the programme has been revised incrementally in line with feedback from students, industry and the accrediting bodies. The programme comprises 24 modules, eight at each level, the majority of which are also taken by building surveying and construction management students. All module descriptors were analysed to try and establish how research was incorporated within the modules and hence the architectural technology programme. Module descriptors were written primarily to address teaching, learning and assessment strategies, there was no explicit reference to research integration within the documentation. In an attempt to gather more data, each module leader was approached and asked to talk about their lecture programme and the extent to which research was integrated into their module.

At level 1, two of the eight module leaders claimed to include research in their lectures, this included their own research findings and research published in refereed journals. This had not been recognized by the students when interviewed earlier. At level 2, only one of the module leaders actively incorporated research findings. Although this individual was not research active, he regarded this as scholarly activity and something all academics should do. Again, this was not recognized by the students. At level 3, six of the module leaders claimed to incorporate research findings into their material, both their own research findings and that of others in their field. There was some acknowledgement of this by the students when interviewed, with four of the six identified by the students as research active.

With the exception of level 2 modules, there appears to be a natural development through the course towards greater research integration within the modules. This is not identified within the course documentation and is difficult to track through auditing procedures. It was also difficult to find much evidence that research is considered within the official feedback mechanisms in place at the university. Module leader reports, student feedback and course committee minutes covering a four-year period were analysed. There was no mention of research in the module leaders' reports or the student feedback forms. Similarly, it was difficult to see a link between teaching and research in the documentation submitted for the RAE.

MONITORING A MODULE

During the monitoring period, the architectural technology curriculum was revised, which led to the design and implementation of a new level 3 module. This aimed to integrate architectural technology, design and management and also to include more of the department's research activity. It took 12 months to develop the module and get it formally approved. The module was subsequently monitored over three academic years. The new module was structured with keynote lectures from research-active staff and project-based coursework. Emphasis was on architectural detailing, functionality and sustainable/ecological issues. Problems with resources, especially the lack of time, appeared to result in missed opportunities to integrate teaching and research. Staff that had been allocated to teach on the module were reallocated by line managers and so the module ran with less resources than originally planned, resulting in fewer keynote lectures. The monitoring process highlighted the increased incorporation of the department's research in the curriculum. In practice, the amount of research actually incorporated in the module delivery was less than that implied in the module descriptor.

During the first year of delivery, feedback was obtained using the university's standard student evaluation forms, but subsequent analysis found that the proforma was not rigorous enough to provide reliable data. A questionnaire survey was designed and issued towards the end of the semester to all students enrolled on the module. This included architectural technology and building surveying students on full- and part-time modes. A total of 67 students completed the questionnaire, giving a response rate of 72%. The feedback on the module was very positive, with students rating the module highest of all level 3 modules for applicability to their chosen profession, ability to learn and satisfaction with subject content. 40% claimed to be aware of research integration within the module, 32% indicated that they were not aware and 28% claimed to be unsure. From the written comments on the questionnaire it was clear that the students had recognized where the research input had been made in the module. Students who were aware of the research input also recommended more research be included in future years. The data supported the findings from the earlier interviews.

DISCUSSION AND CONCLUSIONS

It is not possible to make any claims as to what extent the findings are representative of a larger sample; however, the evidence would tend to suggest that the work is not unrepresentative. Interviews with course leaders at other institutions showed a striking similarity in their concerns and perceived constraints as the course leaders at the case study institution. The observation reported here was specific to one institution and cannot be seen to be representative of a wider sample; however, the observation was useful in helping to illustrate some of the challenges associated with integrating research and teaching.

One of the main issues to emerge was the *ad hoc* connection between research and teaching, as identified by Barnett (1992). This was reflected in academics taking a personal approach to research and teaching integration, a characteristic recognized by the students. The teaching-led culture of the university was reflected in poor support for research at policy level and on a practical level. Although the line managers recognized the link between research and teaching, there was no incentive for them to pursue it at department level, instead it was left to individuals to pursue their own agenda through their module delivery, which they did.

Communication within the department was poor, as was communication between students and teachers. There was no evidence of any formal mechanisms for communication of research results within the department. Dissemination appeared to be entirely at the discretion of individual academics. Individual module leaders made research culture and activity visible to students, but even where this was explicit, students did not always recognize it. The influence of professional bodies was recognized and seen to be becoming more important. Pressure on resources pointed towards less, not greater, integration. From a researcher's perspective, there appeared to be considerable scope for making simple yet practical improvements without impacting on existing resources.

Despite these barriers, there was some evidence of research integration. The architectural technology programme integrated research through some modules, primarily at level 3. This was not articulated in the course documentation and could only be found through interviews and subsequent monitoring. The new module encouraged research-based learning, which the students had found to be highly enjoyable and relevant to their chosen field of study. Academics who had articulated the link between teaching and research to the students were perceived as better teachers than their peers. Students claimed to appreciate those academics that attempted to integrate research into their teaching, which supported earlier research. This research found a positive link between research integration and the quality of the student experience; however, more research is needed to see how research aids and/or improves teaching. The findings support earlier research, which claimed a positive link between research and the learning experience. However, the link was found to be both subtle and not well articulated by academics. The struggle with finding an appropriate balance raised by Fukami (1997) was very evident during the data collection exercise.

RECOMMENDATIONS

Based on the findings reported here, it is possible to tentatively suggest a number of recommendations for improving the synergy between research and teaching. With a little strategic vision, all of the recommendations suggested could be implemented without the need for additional resources. In addition to trying to improve communications at all levels, the following recommendations are made:

- Course leaders (and colleagues) should discuss research with students, especially in relation to how it may benefit their learning experience. Course teams could take a closer look at the undergraduate degree structure and the detailed delivery pattern to identify strengths and weaknesses.
- Managers need to take a wider view of teaching and research instead of seeing the two activities as being unrelated. It is through such a strategy that resources may be better allocated and the scope for an enhanced student experience could be realized. Managers could also make more of an effort to encourage staff to link their research (and/or the research of their colleagues) and their teaching for the benefit of the students.
- The university needs to provide clear and unambiguous direction to line managers and academic staff about the status of research and then resource it accordingly.
- Professional institutions need to give clear direction to course leaders about the level of research incorporation they expect within accredited programmes.

FUTURE RESEARCH DIRECTIONS

The work presented here provides a snapshot of the architectural technology degree and the incorporation of research into the curriculum. It represents the views and opinions of a select number of individuals at a particular point in time. Given the relative newness of the discipline as an undergraduate subject, it would be useful to repeat the approach adopted here in, say, five years' time. This will have allowed time for the undergraduate programmes to mature and comparisons could then be drawn between the findings reported here and those of new studies. Additionally, the methodology employed here could be repeated in other institutions with a view to developing comparative studies. What is evident is the need for more research into curricula design and the role that research/teaching integration plays in architectural technology programmes. In particular, there needs to be a much more detailed monitoring of module delivery to see what is happening in practice.

ACKNOWLEDGEMENTS

The author would like to acknowledge the kind assistance of staff, students and management who were willing to allow qualitative research to be both carried out and the findings disseminated.

AUTHOR CONTACT DETAILS

Stephen Emmitt: Section for Planning and Management of Building Processes, Department of Civil Engineering, Technical University of Denmark, DK-2800 Kgs. Lyngby, Denmark.
Tel: +45 45 25 1660, fax: +45 45 88 5582, e-mail: se@byg.dtu.dk

REFERENCES

Advisory Board for the Research Councils, 1987, *A Strategy for the Science Base*, London, HMSO.

Aldrich, H.E., 1997, 'My career as a teacher' in R. Andre and P.J. Frost (eds), *Researchers Hooked on Teaching: Noted Scholars Discuss the Synergies of Teaching and Research*, Thousand Oaks, California, Sage Publications.

Allee, V., 1997, *The Knowledge Evolution*, Boston, MA, Butterworth-Heinemann.

Astin, A.W., 1993, *What matters in College? Four Critical Years Revisited*, San Francisco, Jossey-Bass.

Ball, C., 1989, 'The problem of research', in *Higher Education Quarterly*, 43(3), 207–215.

Ball, C., 1992, 'Teaching and research', in T.G. Whiston and R.L. Geiger (eds), *Research and Higher Education: The United Kingdom and the United States*, Buckingham, The Society for Research into Higher Education and Open University Press.

Barnett, R., 1990, *The Idea of Higher Education*, Buckingham, The Society for Research into Higher Education and Open University Press.

Barnett, R., 1992, *Improving Higher Education*, Buckingham, The Society for Research into Higher Education and Open University Press.

BIAT, 2000, *Guidance on Submitting Honours Degree Courses for Accreditation by British Institute of Architectural Technologists*, London, BIAT.

BIAT, 2004, *Your Career as an Architectural Technologist*, London, British Institute of Architectural Technologists.

Brew, A. and Boud, D., 1995, 'Research and learning in higher education' in B. Smith and S. Brown (eds), *Research, Teaching and Learning in Higher Education*, London, Kogan Page.

Brookfield, E., Emmitt, S., Hill, R. and Scaysbrook, S., 2004, 'The architectural technologist's role in linking lean design with lean construction', in S. Bertelsen and C. Formoso (eds) *Proceedings of 12th Annual Conference on Lean Construction (IGLC), Helsingør, Denmark, August 3–5*, Helsingør, Denmark, International Group for Lean Construction, 375–387.

Brown, M., 1996, 'Innovation and research – the heart of architectural technology', *Education and Training Forum*, BIAT, London, November (unpublished paper).

Brown, M., 2000, 'Research – the heart of architectural technology', in *Towards a Common Understanding of Architectural Technology*, BIAT, London, February, invited conference paper (unpublished), British Library.

Cohen, L. and Manion, L., 1994, *Research Methods in Education*, 4th edn, London, Routledge.

Colbeck, C.C., 1998, 'Merging in a seamless blend', in *The Journal of Higher Education*, 69(6), 647–671.

Crinson, M. and Lubbock, J., 1994, *Architecture – Art or Profession?: Three Hundred Years of Architectural Education in Britain*, Manchester, Manchester University Press.

Egan, J., 1998, *Rethinking Construction*, London, Department of the Environment, Transport and the Regions.

Emmitt, S., 2000, 'Informing the informers: current research and incorporation into the curriculum', in *Towards a Common Understanding of Architectural Technology*, BIAT, London, February, invited conference paper (unpublished), British Library.

Emmitt, S., 2001, *Architectural Technology*, Oxford, Blackwell Science.

Feldman, K.A., 1987, 'Research productivity and scholarly accomplishment of college teachers as related to their instructional effectiveness: a review and exploration', in *Research in Higher Education*, 26(3), 227–298.

Frost, P.J., 1997, 'Learning to teach' in R. Andre and P.J. Frost (eds), *Researchers Hooked on Teaching: Noted Scholars Discuss the Synergies of Teaching and Research*, Thousand Oaks, California, Sage Publications.

Fukami, C.V., 1997, 'Struggling with balance' in R. Andre and P.J. Frost (eds), *Researchers Hooked on Teaching: Noted Scholars Discuss the Synergies of Teaching and Research*, Thousand Oaks, California, Sage Publications.

Garnett, D. and Holmes, R., 1995, 'Research, teaching and learning: a symbiotic relationship' in B. Smith and S. Brown (eds), *Research, Teaching and Learning in Higher Education*, London, Kogan Page, 49–57.

Gill, J. and Johnson, P., 1997, *Research Methods for Managers*, 2nd edn, London, Paul Chapman Publishing.

Grew, G., 1992, 'Research and the quality of degree teaching – with special reference to consumer and leisure studies degree courses', in *CNAA Project Report 38*, October, CNAA, cited in B. Smith and S. Brown (eds), *Research, Teaching and Learning in Higher Education*, London, Kogan Page.

Harris, I., 1995, 'Research-related staff development: an approach', in B. Smith and S. Brown (eds), *Research, Teaching and Learning in Higher Education*, London, Kogan Page, 102–107.

Jacob, M. and Hellstrom, T. (eds), 2000, *The Future of Knowledge Production in the Academy*, Buckingham, The Society of Research into Higher Education and Open University Press.

Jenkins, A.J., Blackman, T., Lindsay, R. and Paton-Saltzberg, R., 1998, 'Teaching and research: student perceptions and policy implications', in *Studies in Higher Education*, 23(2), 127–141.

Kogan, M. and Henkel, M., 1992, 'Constraints on the individual researcher' in T.G. Whiston and R.L. Geiger (eds), *Research and Higher Education: The United Kingdom and the United States*, Buckingham, The Society for Research into Higher Education and Open University Press.

Latham, M., 1994, *Constructing the Team*, London, HMSO.

Lindsay, R., 1998, 'Teaching and research: the missing link', in *Teaching Forum*, 45, 9–11.

Neumann, R., 1994, 'The teaching-research nexus: applying a framework to university students' learning experiences', in *European Journal of Education*, 29(3), 323–338.

Nord, W., 1996, 'Research/teaching boundaries', in P.J. Frost and M.S. Taylor (eds), *Rhythms of Academic Life*, Newbury Park, California, Sage.

Pascarella, E.T. and Terenzini, P.T., 1991, *How College Affects Students*, San Francisco, Jossey Bass.

Perkins, J., 1998, 'Integrating research and teaching', in *Teaching Forum*, 45, 3–4.

Potter, J., 1996, *An Analysis of Thinking and Research about Qualitative Methods*, Mahwah, NJ, LEA.

QAA, 2000, *Academic Standards: Architectural Technology*, May, Gloucester, Quality Assurance Agency for Higher Education.

Race, P., 1995, 'Competent research: running brook or stagnant pool?', in B. Smith and S. Brown (eds), *Research, Teaching and Learning in Higher Education*, London, Kogan Page, 75–88.

Ramsden, P., 1998, *Learning to Lead in Higher Education*, London, Routledge.

Ramsden, P. and Moses, I., 1992, 'Associations between research and teaching in Australian higher education', in *Higher Education*, 23(3), 273–295.

Rosen, M., 1991, 'Coming to terms with the field: understanding and doing organisational ethnography', in *Journal of Management Studies*, 28(1), 1–24.

RIBA, 1962, *The Architect and His Office*, London, Royal Institute of British Architects.

Seale, C., 1999, *The Quality of Qualitative Research*, London, Sage Publications.

Smith, B. and Brown, S. (eds), 1995, *Research, Teaching and Learning in Higher Education*, London, Kogan Page.

SAAT, 1984, *Architectural Technology: The Constructive Link*, London, Society of Architectural and Associated Technicians.

Symon, G. and Cassell, C. (eds), 1998, *Qualitative Methods and Analysis in Organisational Research: A Practical Guide*, London, Sage Publications.

Utley, A., 2001, 'Research makes you teach better', in *Times Higher Education Supplement*, 23 March, 5.

Yeomans, D.T., 1996, 'Design PhDs in architecture', in *Doctorates in Design and Architecture, Proceedings of European Association for Architectural Education, Volume 1, Delft University of Technology*, 10–12 February, Delft, EAAE, 118–123.

ARTICLE

The Effectiveness of E-learning

Ezekiel Chinyio and Nick Morton

Abstract

Our focus is on the effectiveness of e-learning, reflecting on our experiences both as tutees and tutors. The paper considers these two strands, culminating in an overview of the enablers and inhibitors of this learning medium. In the first strand, an ethnographic study is discussed, involving 36 students on a Post Graduate Certificate of Education (PGCE) programme. The major assessment of this course involved the use of an e-learning environment known as Moodle. An e-conference by means of Moodle was used to facilitate group discussions and to inform assignment writing. On reflection, the approach proved effective for both communication and comprehension. It circumvented the potential problems of group meetings and helped overcome other common communication difficulties. However, conversely, it also suggested that a drawback of e-conferencing might be the absence of nuances in speech and body language. In the second strand, the paper considers the *delivery* of e-learning at UCE Birmingham and in particular the built environment subject area. This is based on the authors' participation as tutors. The role of Moodle as a tool in delivering effective teaching is addressed specifically, outlining some of the successes and glitches encountered so far. Using a student satisfaction survey and interviews with staff members who have used Moodle, the authors reflect on some of the inhibitors that slow the uptake of e-learning as a broader pedagogical approach. Workshops, one-to-one discussions and induction programmes are seen as methods of facilitating the broader uptake of e-learning.

■ *Keywords* – **E-learning; conference; higher education; Moodle; communication; reflection**

INTRODUCTION

Learning is a lifelong phenomenon. Institutions exist to enhance learning in a structured and guided manner; for example, in higher education (HE) institutions, learning is typically organized according to courses and each course consists of several modules or subjects of study. The completion of each one usually requires the individuals involved to attain certain learning outcomes upon which they can be assessed.

However, individuals attach different preferences to different learning methods. For instance, people who are left-brain dominant tend to prefer text and numbers, whereas those that are right-brain dominant tend to prefer graphics. The dynamics of human cognition is a complex subject. It is thus incumbent on a tutor to identify and use the learning approach that fits the tutee(s) best, a process that requires careful consideration.

Indeed, there is a considerable body of research into different learning approaches (see for example, Coles, 1997; Cassidy, 2004). They are frequently characterized as either *deep* or *surface* approaches, representing two extremes of a continuum (Marton and Säljö, 1976a,b; Biggs, 1999). A surface approach places greater emphasis on text, and is characterized by simple memorization and repetition of information. A deep approach, meanwhile, emphasizes the derivation of meaning and the understanding of context and interrelationships. Significantly, there is a correlation between deep approaches to learning and high-quality outcomes (Morón-Garcia, 2004). Therefore, the deeper approaches to learning are greatly encouraged.

Sometimes a deep approach may not seem feasible; for example, where a great deal of factual material has to be conveyed, or the students are making a dramatic change in their learning subject or style, what might be dismissed as constituting a surface approach may in fact be entirely necessary information transfer. Even so, multi-dimensional learning is still possible. Many learning models emphasize the importance of dialogue (for example, the conversational framework developed by Laurillard (2002)). Educational research has shown that more effective learning takes place if learners are actively engaged, as opposed to being passive listeners (Webb *et al*, 2004). This has brought a move towards a greater level of 'student-centred' and experiential learning, the latter conceptualized by Kolb (1984) as an ongoing cycle of active experimentation, concrete experience, reflective observation and abstract conceptualization. Student-centred learning is flexible and involves supplementing a traditional 'one-way, one-venue' teaching style with activities that are more varied in terms of the place, pace, time and content (Race, 1998) and promote an active dialogue between tutor and tutee (Laurillard, 2002). Examples of flexible learning activities that facilitate dialogue include self-guided field trips, student-led projects, seminars, workbooks and e-learning (SLICE, 2002).

E-learning, the particular subject matter of this paper, has been defined in many different ways. The broad conceptualization we shall use is that favoured in many recent UK governmental strategy documents, i.e. simply, the use of ICT in learning opportunities (DfES, 2003, 2005; HEFCE, 2005). One of the principal advantages of e-learning is the flexibility ICT offers in terms of pace and distribution of learning (Macpherson *et al*, 2004), achieved without time and place restrictions (Galagan, 2000). It encompasses many specific ICT-delivered learning and teaching activities, such as e-tutoring (supporting and assisting students online; see Duggleby *et al*, 2002) and e-conferencing (mediating communication through online discussion forums; see Fisher, 2004). This paper will address applications of these activities in an HE environment.

The flexibility inherent in e-learning accords well with the differences between learners, in that they can learn at a time and pace to suit their own capability and circumstances (Sandelands and Wills, 1996; Caudron, 1999). Flexibility also enables students who get tired, taxed or bored to put their learning material aside until they are ready to resume. It further provides opportunities for off-campus learning, a mode preferred by many mature and part-time students (Murray *et al*, 2004). In the past, tutors had controlled the pace, place, time and style of learning. However, where e-learning is used, the control of these elements has largely been transferred to the learners (Blass and Davis, 2003), who can now be empowered to undertake study activities that might be off-campus or carried out without total reliance on lecturer contact (Race, 1998, 2001). When students are accorded more flexibility, they are more likely to become engaged and to develop more autonomy and improved lifelong learning skills.

Whichever learning method is adopted, a measure of its success is its effectiveness. In one-to-one (tutor-learner) situations, the learning method can be chosen easily in accordance with the preference of the learner, whereas in one-to-many situations, the preferences of the learners may conflict with each other. When such conflicts arise, the effectiveness of learning can be undermined and for groups it is often worthwhile using a combination of learning methods dependent on the needs of that particular group. In so doing, the tutor is most likely to achieve an overall learning approach that is both inclusive and, centrally to the theme of this paper, effective (SLICE, 2002).

It should be noted that this paper will not investigate the cost implications of e-learning either in the short or long term. There is a substantial body of literature on this contentious issue (Laurillard, 2001; Blass and Davis, 2003; Littlejohn and Higgison, 2003) and this paper assumes that such a consideration would have been made during the feasibility stage of designing an e-learning system. What the paper concentrates on is the efficacy of e-learning, assuming that a decision to use such an approach has been made.

RESEARCH METHODOLOGY

In the following sections, a study is described where the effectiveness of e-learning was investigated. The study was carried out at the University of Central England in Birmingham (UCE Birmingham) over a period of 12 months. The overarching methodological approach employed was that of a case study (Yin, 2003) informed by a qualitative and quantitative survey of staff and students. There were two parts to this study, each lasting six months, but the two periods did not overlap with each other. Figure 1 illustrates the timeframe of the survey.

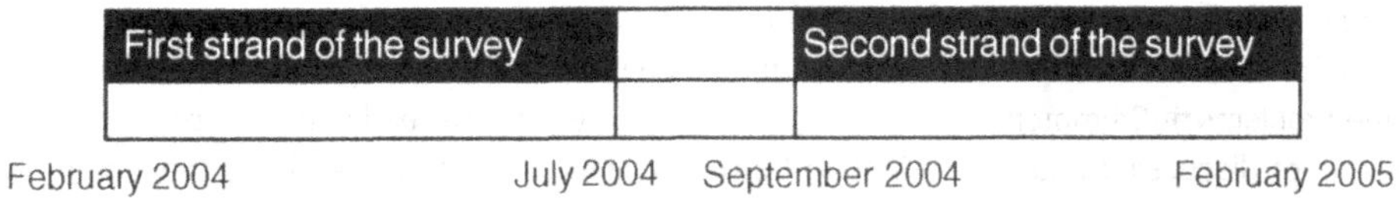

First strand of the survey		Second strand of the survey

February 2004 July 2004 September 2004 February 2005

FIGURE 1 **Research timetable**

FIRST STRAND: MOODLE AS A MEANS OF EFFECTIVE COMMUNICATION

The first strand of the study was carried out by means of a qualitative survey. The sample consisted of 36 members of staff of UCE Birmingham on a Post Graduate Certificate of Education (PGCE) course. The authors were part of this group; as such, there was an ethnographic connotation to this study because data were collected partially by participant observation (Creswell, 1994).

The PGCE course in question covered four major learning subjects, each of which was assessed. The tutees (i.e. our sample) attended 12 full days of learning in the 2003/04 academic year. The course was designed to operate on a model of *hybrid* or *blended learning* (Littlejohn and Higgison, 2003; Thorne, 2003), where traditional classroom teaching was supplemented, but not replaced, with other approaches such as workshops and e-learning. Part of the programme was an exercise in peer-assisted learning, and for this section the Moodle e-learning environment was utilized.

Moodle is a virtual learning environment – perhaps better described as a course management system – available as open source software to any interested institution or organization. In active use in at least 114 countries across the world (Moodle Sites, 2005), it is a Web-based tool for knowledge exchange with facilities allowing synchronous and asynchronous contact, group discussion boards, video clips and participant profiling, as well as functioning as a more typical document repository.

The e-learning approach was designed to encourage course participants to communicate with each other more and to reinforce learning in a deeper way. In particular, given our differing job commitments, e-conferencing was the ideal way to communicate effectively outside the attendance days.

The PGCE course tutors introduced Moodle to the students and demonstrated how it worked. The students were encouraged to access the system over a trial period lasting about six weeks, after which the tutors used the discussion forum feature to initiate a discussion on the efficacy of learning. Students then joined in this debate, gradually increasing their interaction with the new system.

In due course, the tutors introduced wider discussion themes to the tutees. These were related to the learning outcomes of the course through an assignment concerning a hypothetical HE institution called Crumpton (MacKenzie and Staley, 2000; Staley and MacKenzie, 2001). Information on Crumpton was provided in the form of video clips and written documents. This information was displayed and explained to the students and placed on Moodle for continuous access. Students were then asked to discuss the problems at the institution and to explore ways of solving them, initially through a group discussion in Moodle and ultimately through a report by each course participant.

The tutors ensured that Moodle was available and accessible to all tutees by offering assistance with the system when necessary. Although they initially acted as facilitators in the discussion, setting the first questions in each theme and sometimes responding to comments posted in response, they thereafter allowed the e-conference to be self-moderated by the students.

The guidelines for each participant were to:

- participate in the discussion of each theme
- make a minimum of three contributions to each discussion theme; and
- develop reference material for the assignment based on the discussions.

In addition, students were free to initiate relevant themes of discussion within the broad subject of learning. For practical reasons, the e-conference was limited to a timeframe of nine weeks and the discussion on each of the themes was timed to close on a given day. The tutors posted reminders of the need to participate in the on-going conference and also to give advance warning of any impending deadlines. At the end of the e-conference, each student produced an individual report that addressed the problems at Crumpton.

As a means of discussing and disseminating both outside experiences and the information received through lectures, seminars, group discussions, workshops and the formation of active learning sets, e-conferencing thus provided a major blended learning forum within the PGCE programme. As noted by Wilson and Stacey (2004: 33), 'using group conferences as a central communication space provides a means of enabling the groups to socially construct knowledge'.

In analysing this process during the writing of this paper, it was of course useful that all communications had been captured and saved through Moodle. In addition, the authors kept records of proceedings of the PGCE course. Finally, on its completion, the authors were able to reflect on their experiences in order to extract lessons that could be learnt.

Given the qualitative nature of the study, a thematic synthesis of the data was employed. Essentially, we sought to identify good (positive) aspects that may enhance the effectiveness of e-learning, and downsides (negatives) that may inhibit either uptake or operation. Thus, themes were identified in the data and allocated to these two broad categories. The process was reviewed iteratively (Leedy, 1997) as we reflected on the data and the PGCE experience, in line with the model of reflective study proposed by Boud *et al* (1985).

SECOND STRAND: MOODLE AS A TOOL IN ENABLING THE DELIVERY OF EFFECTIVE TEACHING

The second part of this study was based on a triangulated survey of staff and students' use of e-learning, this time reflecting the authors' experience as tutors. The context for this was the adoption of Moodle in June 2004 as UCE Birmingham's first institution-wide e-learning system. In common with other faculties across the university, Built Environment (BE) was encouraged to begin experimenting with the development of online content prior to the commencement of pilot modules in September 2004. BE was, however, one of the first three faculties to be awarded a part-time post to facilitate this experimentation process, appointed as a secondee to the central Learning Technology Development Unit. One of the authors was chosen to be this so-called Moodler with responsibility for the three schools comprising the BE faculty.

One of the early, but ongoing, tasks of these facilitators was to disseminate the idea of Moodle as a teaching innovation that would make a positive contribution to the student experience and to the quality of teaching. To this end, 15 BE staff were surveyed regarding their interest in various aspects of online delivery (Figure 2). The aspiration towards greater use of technology and student interactivity is evident from these results; the challenge was how to translate this interest into reality.

Colleagues interested in developing online content have been encouraged to view Moodle as being most effective when used as an innovative and interactive learning support system rather than simply as a passive repository for lecture notes. The BE modules supported by Moodle make a full range of online features available to students, including online course documents and reading lists, Internet links, PowerPoint lecture notes, discussion forums, personal reflective journal space, group assignments and the facility to upload draft papers for comment and discussion with tutors.

Commensurate with the increasing number of live modules, the subsequent growth of registered users of the university-wide system has been impressive. Figure 3 provides an overview of the progressive uptake.

This paper addresses the results of surveys conducted with both staff and student users of Moodle, aiming to assess the effectiveness of the system and review progress of its implementation. These satisfaction surveys were carried out within BE and across the university towards the end of the first semester of Moodle operation (January 2005).

The student survey consisted of a questionnaire, asking respondents to assess three clear-cut statements and rank them on a five-point Likert scale from 'strongly agree' to 'strongly disagree':

- 'I enjoy using Moodle'
- 'Moodle has helped me learn the subject'
- 'I would like more modules to be supported by Moodle'.

The survey form also gave space for individual qualitative comments. In all, 338 UCE Birmingham

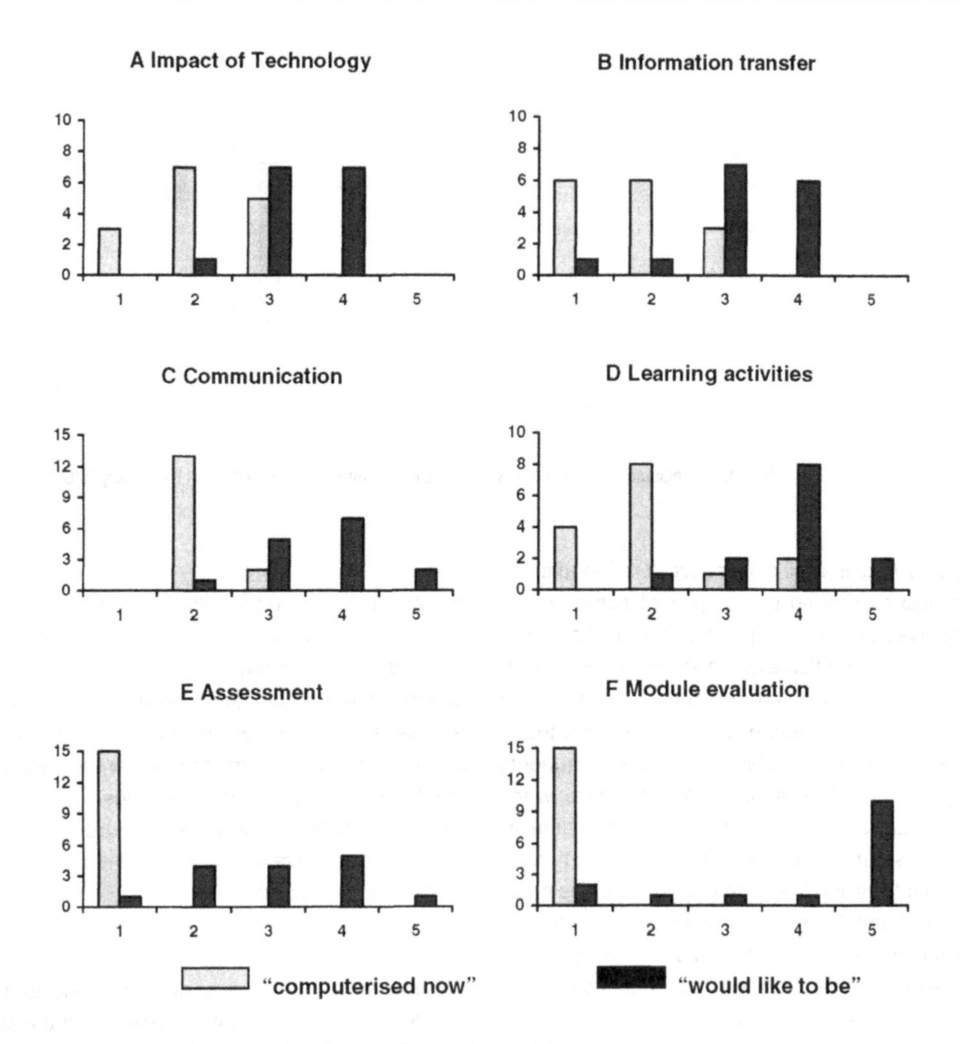

FIGURE 2 e-delivery: scales of impact of electronics on delivery

students completed this survey. A statistical analysis was used to gauge the students' assessment of e-learning via Moodle, while their qualitative comments were identified and compiled, as in the first strand, into the two categories of positives and negatives.

In addition, six BE staff members who had started to utilize Moodle for teaching were interviewed on their experiences. These interviews were held at various times in the period from September 2004 to January 2005. In line with the other data, the information generated was broadly categorized by theme and classified as positive or negative.

FINDINGS: EVALUATING THE UCE BIRMINGHAM EXPERIENCE

The discussions in the following sections are informed by the results of both the staff and student surveys. The findings are presented in accordance with the two strands of the research.

FIRST STRAND: TUTEE'S REFLECTIONS ON E-LEARNING

Interim feedback revealed that some participants felt they were initially slow to engage with the e-conference as a mode of communication, even though they

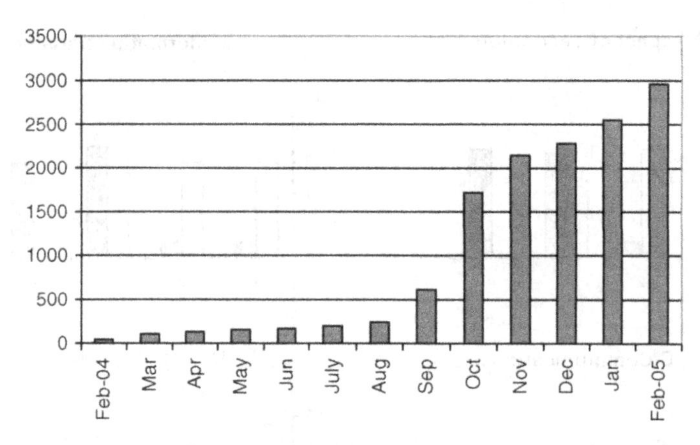

FIGURE 3 **Number of registered Moodle users at UCE Birmingham, February 2004 – 11 February 2005**

recognized its value in easing their access to Moodle and e-learning. Busy workload on the part of some course participants was cited as a major reason for this sluggish start. Some other participants attributed their lack of enthusiasm to uncertainty concerning the unknown. At the front-end of the e-conference, therefore, the tutors had to use persuasive reminders to prod some students into using Moodle. This is very often the case when contributions to discussions are initially made voluntarily. Introducing assessment is a considerable motivation for participation (Staley and MacKenzie, 2001) and this could be done from the beginning if a quicker result was desired. Momentum was maintained once everyone was active in the discussion; however, since there had been a slow start, some students could not make the necessary minimum contributions to the discussions. This resulted in a three-week extension being made before the e-conference was drawn to a close.

Positives

Ease of communication

Once this momentum was achieved, it was generally felt that the e-learning approach proved very effective for communication and ease of comprehension. The communication was asynchronous, i.e. there could be a time lag between sending, receiving and replying to any given communicative event (Webb *et al*, 2004). The tutees could thus contribute to the conference at their convenience. This circumvented the potential problem of time shortage, which often plagues group meetings.

Efficient communication

The e-learning medium also helped in overcoming personal communication problems, like poor tone of voice, speed of speech, etc. The potential negative aspects of leadership and coordination did not arise because everybody was an equal contributor and no opinion was imposed on the conference participants. Everybody had the opportunity to learn ideas from the conference. It also afforded an opportunity to test out such ideas, because a new line of discussion could be initiated by any member.

User friendliness

The e-learning mode was generally felt to be user friendly, in the sense that no one felt intimidated by either peer or tutor. The risk of making an impulsive statement and regretting it later was also eliminated in the Moodle environment. A 30-minute time lag before any contribution was communicated to others allowed participants a second chance at editing and thus facilitated more effective communication.

Interactivity

The ability to utilize humour was not undermined as contributors had their free choice of words. Each contributor could also respond at his/her own speed and style, giving them more effective control. No one was monopolizing the group, as no one was in charge; participants enjoyed the freedom to be themselves.

Enhanced communication

Occasionally, one may come across those who have difficulty with communication through attributes such as stammering, different accents, etc. These can cause problems in group discussions, but the e-communication channel circumvented any such issues.

Enabling blended learning

Because the e-conference was used as one among several means of learning on the PGCE programme, it was felt to have enhanced the overall experience of the course. Our study suggests that e-learning can be an invaluable tool in conjunction with other pedagogical approaches such as groupwork and action learning sets. This supports the recent wealth of material concerning blended learning as a useful pedagogical approach (Ellis and Calvo, 2004; Koos and Betty, 2004; Alonso *et al*, 2005).

Freedom and control in participative learning

In addition to responding to the tutor-specified discussions, some tutees initiated discussion topics that ran alongside the main conference. Valuable and insightful contributions were made in these supplementary discussions, concerning topics such as project-based learning, dealing with reluctant students, time-controlled assignments and technology in teaching.

Negatives

No face-to-face contact

A possible downside of e-learning is that it eliminates or restricts face-to-face contact. Individual preferences differ. Whereas some people enthusiastically embrace the use of technology, others are uncomfortable with it and, for them, the absence of face-to-face contact will be inhibitive. However, in this case we also met each other physically during the attendance days and, therefore, when communications were received they were not from someone totally unknown. In addition, Moodle enables users to upload a picture (typically a small photograph of themselves) as part of their profile; this image automatically appears alongside any contribution they make and thus personalizes the process. For some, though, this will of course be insufficient to replace live, face-to-face contact.

Absence of verbal communication

Similarly, it is relatively easier to explain an issue at length in a face-to-face situation than in a textual mode. Some people are adept and indeed thrive at public speaking. To such people, innovative ideas are generated and communicated better while speaking, not necessarily while writing. This is another possible inhibitor to effective online communication.

Difficulty of visual representation

The e-conference in our observation was a largely textual medium. Even when images can be uploaded, it is rarely easy to embellish ideas with graphics within the constraints of an online discussion. Those right-brain dominant people who tend to prefer graphics as a mode of communication may find this limiting or frustrating.

Lack of nuances in information

E-conferencing will curtail some further subtleties present in other modes of communication (Linser and Ip, 2002). Aspects of emphasis, de-emphasis, body language, etc. were obviously absent, and this can potentially undermine the efficacy of communication. Some speakers, for example, are adept at using such nuances to supplement their verbal speech. Our survey suggested that some participants did feel uncomfortable without the usual verbal and non-verbal cues received in ordinary conversation.

Individual circumstances

Consideration needs to be given to groups where, for example, dyslexia or visual impairment makes text-based discussion difficult for some members. However, this was not an issue in our survey.

Summary

These drawbacks can inhibit the communication and exchange process, and, therefore, by implication the effectiveness of e-learning. However, the majority of participants in the UCE Birmingham survey felt that they did not significantly affect their experience. Contributors generally felt able to express opinions freely and effectively; helped, no doubt, by the fact that members of staff should hopefully make for mature and thoughtful students. Ultimately every single one was able to complete the e-learning element of their PGCE assignment successfully.

SECOND STRAND: THE E-LEARNING JOURNEY SO FAR

Having been part of a successful cohort ourselves, we were now interested to see if the wider student body of UCE Birmingham would find Moodle similarly engaging. Encouragingly, the response to the student survey was overwhelmingly a positive one, and suggests that Moodle has so far been appreciated by its users. Furthermore, the feedback suggests that of all faculties in the university, BE was one of the most receptive to e-learning – albeit on a sample size small in comparison with some other larger faculties. The mean score on a Likert scale in response to the statement 'I enjoy using Moodle', for example, was 4.33 in BE compared with an average of 3.77 across UCE Birmingham as a whole (see Figure 4 for further results).

Similarly, BE students responded more positively to the statement 'Moodle has helped me learn the subject' than students across the university in general (Figure 5). The UCE Birmingham average of 4.03 (4.42 in BE) indicates that e-learning has not only been effective in simple information transfer, but that it can also enhance students' perception of the learning process. Indeed, Moodle can be seen as a useful framework to position students for deeper learning. As Biggs (1999) suggests, deep learning is more likely to occur where four elements are present – a well-structured knowledge base, a motivational context, learner activity and interaction with others. These are all features that Moodle support can provide.

The most emphatic evidence of student enjoyment in BE came in the response to the question 'I would like more modules to be supported by Moodle'; 10 out of 12 respondents selected the highest category 'strongly agree', and the mean score given to this question was 4.67 compared with a university-wide mean of 4.14. Figure 6 shows responses to this question across the university; not one student in BE responded negatively.

Individual comments were also highly encouraging, for example:

- 'The links to the appropriate sites was [sic] helpful when researching and helped to save time. I found this a big help. It was also good to check module info and would have been excellent to have it for other modules. PLEASE DO IT FOR EVERY MODULE.'
- 'It makes learning a lot more interesting. I like the info/access to other students and the helpful websites.'
- 'This is my first experience in using Moodle as learning section [sic]. I enjoyed it very much.'
- 'It's fantastic. We need more modules available on Moodle.'

The true extent of the impact Moodle makes on teaching and learning will only become clear if a substantive improvement in grades can be

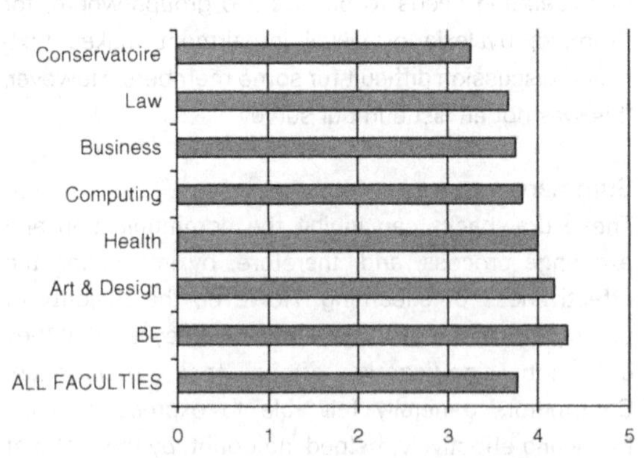

FIGURE 4 **Mean response to the question 'I enjoy using Moodle', shown by faculty**

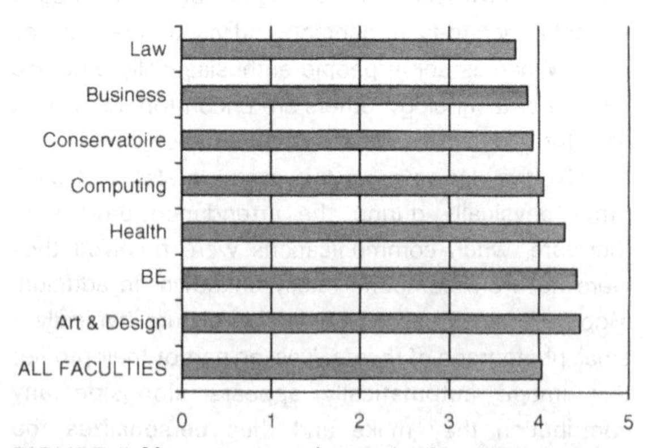

FIGURE 5 **Mean response to the question 'Moodle has helped me learn the subject', shown by faculty**

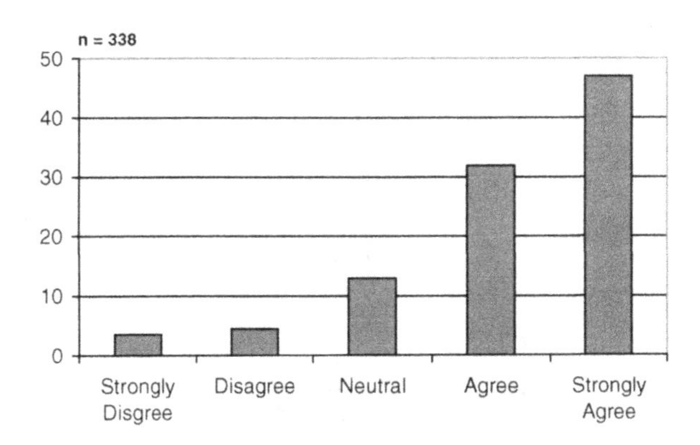

FIGURE 6 **Percentage response to statement 'I would like more modules to be supported by Moodle', across UCE Birmingham**

demonstrated. However, as modules are currently being delivered with Moodle support at UCE Birmingham for the first time, it is too early to carry out such a study. The 'no significant difference phenomenon' (Russell, 1999), frequently quoted in the e-learning debate, suggests that results do not differ much regardless of the medium of teaching; however, other research has suggested that results do improve as a result of online delivery (Fallah and Ubell, 2000; Piccoli *et al*, 2001).

ADOPTING E-LEARNING AS A BROADER PEDAGOGICAL APPROACH

Evaluation of the initial Moodle provision has clearly shown that BE students have, for the most part, found it to be a useful and rewarding experience. However, to achieve effective e-learning across the whole range of BE teaching will require a much broader strategic commitment to embed and integrate Moodle into all courses.

To a large extent, student uptake of Moodle is driven by the content that staff make available; with few exceptions, the BE students at UCE Birmingham have responded positively to its introduction and, it can be assumed, will continue to do so in the future. It would seem that the immediate priority is, therefore, to encourage more colleagues to develop online content in order to meet the demand that appears to exist.

There are a number of potential hindrances that can be characterized as inhibitors of a wider uptake. All have been witnessed to a greater or lesser degree in the authors' experience as tutors or have arisen in discussions with colleagues. They include pressures on staff yet to explore the use of e-learning, but also students' ability to engage with the system – which, after all, will be a determining factor for many considering developing e-learning in their teaching.

Inhibiting factors for staff

Moore's chasm

Here, the model of innovation diffusion proposed by Moore (1991) is a valuable analogy and it is useful for interpreting the experience at UCE Birmingham. Initially developed with reference to the field of marketing, but since applied to the growth of technology in teaching (for example, Geoghegan, 1995), the model suggests that while 'innovators' and 'early adopters' are enthusiastic in their uptake of a new development, the real challenge is to move from a loyal minority of such users to a mainstream majority. It is this chasm that many innovations fail to safely negotiate (Figure 7).

Lack of confidence using a new system and form of delivery

For some colleagues, the prospect of adopting e-learning is a daunting one. For the even mildly technophobic, the thought of committing a whole module to Moodle can be overwhelming, particularly if it is felt they are being coerced into adopting a technology-driven approach they are not predisposed towards.

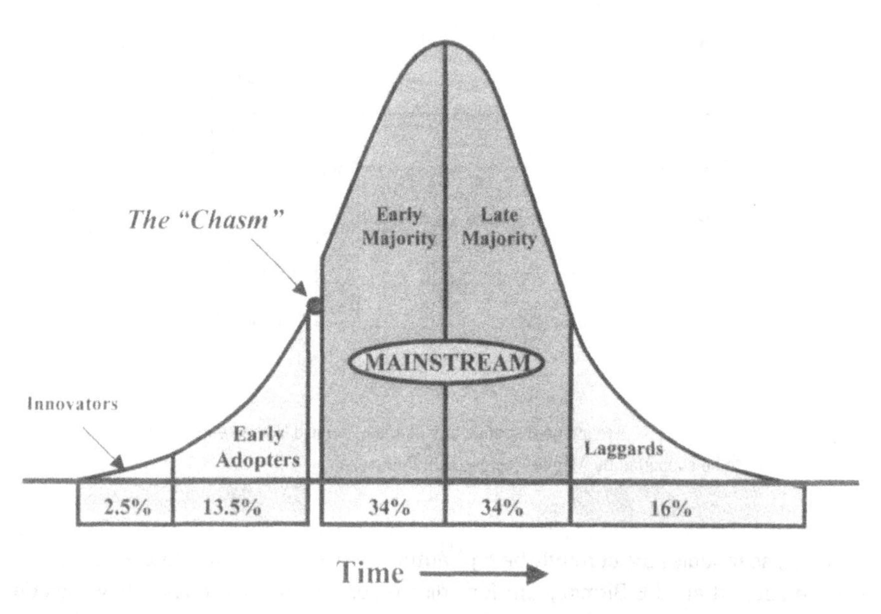

Source: Staley and MacKenzie, 2001: 142

FIGURE 7 Moore's chasm

Lack of time – pressure of work and other commitments

Even where colleagues are less wary of e-learning as a concept, a simple lack of time may prevent them from engaging with it as fully as they might have liked. In the case of the authors' own experience, this has been further complicated by a major restructuring process in the BE faculty which inevitably has left e-learning a rather low priority for some otherwise enthusiastic staff. It is the role of the Moodler facilitator to be a champion of e-learning as well as to strike the right balance – to accentuate the added-value benefits Moodle offers teaching without becoming a nuisance and creating another unwanted pressure on an already crowded day.

Alienation – old habits die hard

Some colleagues may reasonably ask: 'If it already works, why change it?' In this category, one could include staff who have previously used interactive lectures successfully and who, since their present approaches are working well, are not in a hurry to adopt a new and possibly uncertain approach. Indeed, Moore (1991) suggests that the success of early adopters can actually deter the mainstream from following, perhaps because the workload appears intimidating or even because the achievements of others become irritating. These mindset issues can hopefully be managed and resolved with time.

Inhibiting factors for students

Problems of accessibility

Technology, for all its advantages, can sometimes seem to cause as many difficulties as it resolves. Increasing adoption of e-learning raises the expectation that students will have access to the Internet from home, even where adequate access is made available in the university. Not only is this potentially a financial burden, but remote access also brings inevitable technical problems that are beyond the reach of IT support. The UCE Birmingham system registered a surprisingly large number of hits even over the 2004 Christmas holiday period, for which it would obviously have been difficult to provide technical assistance if it had been required.

In the case of the authors' own experience, these unavoidable dangers have been exacerbated this year by local circumstances. The Moodle log-in system is dependent on the student having a university ID generated when the registration process is completed; this, in turn, is often reliant on local education authority funding being confirmed, and where this process has been subject to significant delays, students have been left unable to access online material for sometimes lengthy periods. Obviously, as a tutor, this is particularly frustrating and is as unsatisfactory for staff as it is for the students concerned.

Lack of engagement with online learning/computer use generally

It will inevitably be the case that, for some students, e-learning is simply never going to be an enjoyable experience. While careful tutoring and an investment of time in reassuring individuals who are uncertain of their computing skills will go some way to ameliorate this, it is important not to disadvantage students who engage less successfully with e-learning not through any lack of effort, but simply because their personal approach to learning is a different one (Jones *et al*, 2004).

The continuous engagement of students can also become an issue in e-learning, as some tutees can lose concentration along the line. However, this potential loophole is not peculiar to this approach; even in face-to-face sessions, attendances can drop. In fact, the automated tracking of participation that many online packages offer means that monitoring and control is often made easier.

Leading the way – shyness in making their voice heard

In the first part of this paper, we discussed the positive benefits of communicating online and the equality of opportunity it offers those participating. However, that dealt with our experience among a peer group of colleagues; for first-year undergraduate students, the thought of making a first contribution to a discussion forum, for example, might be almost as daunting as doing so in front of a live class. Many will require specific encouragement and support from the e-tutor.

Bridging the chasm: methods of engagement

In light of Moore's (1991) chasm, the inhibitors outlined above are not totally unexpected. Staley and MacKenzie (2001) suggest that e-learning will not become adopted by the mainstream until staff feel it is personally worthwhile to do so, perhaps through a combination of incentives and motivation with peer support. Our personal experience, and the interviews conducted with BE staff at UCE Birmingham, suggest a number of ways to proceed. These include the holding of workshops, one-to-one discussions and, for students, introductory support sessions.

Staff workshops

Lunchtime workshop sessions have proved a successful method of dissemination within the BE faculty. In conjunction with centrally organized events and initiatives such as a campus-wide 'Moodle Week' and a newsletter highlighting online developments, they have proved popular in publicizing at least the existence of Moodle. Inspiring though they might be for early adopters, however, alone they are unlikely to persuade the silent majority to actually begin to use the system; indeed, it is likely to be those with some initial enthusiasm that attend such workshops anyway. One method of increasing the reach of workshops is to offer different types of session of variable lengths and at different times (Wiles and Littlejohn, 2003).

A peer support culture – using one-to-one sessions

Working with colleagues on an individual basis is time-consuming but irreplaceable in terms of what can be achieved. In the same way that a tutorial can build confidence and skills in a student, so investing time with individual members of staff can demystify the development process and gradually foster independent skills. The whole ethos of the Moodler role is to act as an enabler – to give staff the confidence and ability to use the system for themselves. The efforts of the BE Moodler are beginning to be repaid with an increased acceptance of Moodle across the faculty. This reflects the findings of similar cases discussed by Wilson and Stacey (2004), where peer support based on establishing good practice has proved a useful method of encouraging the wider use of e-learning.

Being responsive to changing needs

It is important to remember when designing such a support system that the needs of staff will vary according to confidence and ability, and that these

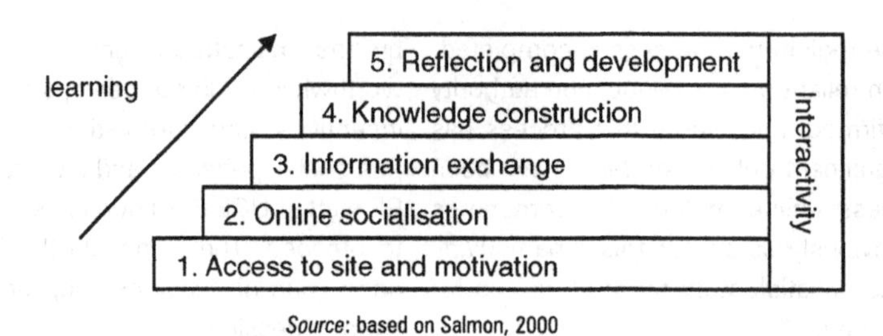

Source: based on Salmon, 2000

FIGURE 8 Stages in online knowledge construction

needs will change over time. Staff requirements will initially be determined by competency in using e-learning, ranging from novices interested in basic 'show and tell' sessions, to experts who act as role models and actively pursue research and development (Bennett *et al*, 1999; Wilson and Stacey, 2004).

Introductory sessions for students – creating an online community

Embedding e-learning into first-semester, first-year modules introduces students to what will hopefully become a recurrent theme of their time at the HE institution. In the case of UCE Birmingham, it is sensible, therefore, to register students on to the system and begin using Moodle from the first opportunity – typically as part of the induction process. In this initial contact, Moodle can function primarily as a social medium, a method of posting profiles and notices, chatting and exchanging details (something that contemporary undergraduate students are increasingly likely to be familiar with from previous online experience).

Other research into e-learning has also suggested the importance of creating communities online (Palloff and Pratt, 1999) and that social presence can be enhanced by making initial postings mandatory and by using conversational style (including 'emotions') in all online communications (Hostetter and Busch, forthcoming). This socialization or bonding process has the added advantage of helping to break down barriers to the use of computing generally – an example of so-called 'pleasurable learning' (NESTA Futurelab, 2004). Introductory sessions thus guide students through the first two stages of a five-stage model (Salmon, 2000; see also Skinner, 2004) for developing deeper learning and knowledge construction skills (Figure 8).

By gaining in confidence through a socialization process and beginning this journey towards deeper learning, students will hopefully develop the characteristics suggested by Palloff and Pratt (2003) as ideal for participants in an online course – openness, flexibility and humour, honesty and the willingness to work collaboratively.

CONCLUSION

Our experience, corroborated by the findings of staff and student surveys, firstly indicates that e-learning efficiently facilitates information transfer. Furthermore, the use of Moodle assists effective communication and, ultimately, will position students for deeper learning. It suits the styles and speeds of different learners, as well as empowering them to take control of their learning. At the same time, it affords tutors greater opportunity to intervene in the process where necessary – for example, to provide prompts to students to encourage interaction. Therefore e-learning can be seen to be a successful medium for delivering both effective learning and effective teaching.

However, the human dimension to e-learning must be monitored, especially in the initial stages. Some learners (and, indeed, staff) could be timid or unenthusiastic about learning via technology and will need to be motivated to embrace it. Regardless of the individual circumstances of a course or institution, it is also inevitable that obstacles to the widespread implementation of e-learning will be encountered. Nevertheless, it is clear that the effort required to overcome such difficulties is usually a worthwhile one. In many cases, given the level of contemporary technological usage, tutors and tutees should not find it too hard to adapt to e-learning. The e-learning curve can be very short indeed, after which knowledge exchange can be very effective; the BE students surveyed in this case study have certainly proved receptive to the idea of e-learning and are keen to see its expansion in the future.

It is important to stress that e-learning is a supplement to, and an enhancement of, regular classroom teaching, and not a substitute for it. It is imperative to reassure students, particularly those who are less inclined to enjoy using computers in general, that face-to-face contact and teaching time will not be undermined by the use of Moodle or any other e-learning medium.

ACKNOWLEDGEMENTS

The authors wish to thank the tutors of the Staff and Student Development Department of UCE Birmingham, and in particular Alan Staley, for support given with the data analysis and surveys contained in this paper.

AUTHOR CONTACT DETAILS

Ezekiel Chinyio and Nick Morton: School of Property, Construction and Planning, UCE Birmingham, Perry Barr, Birmingham, B42 2SU, UK. Tel: +44 (0) 121 331 5631/5145, fax: +44 (0) 121 331 5172,
e-mail: ezekiel.chinyio@uce.ac.uk, nick.morton@uce.ac.uk

REFERENCES

Alonso, F., López, G., Manrique, D. and Viñes, J.-M., 2005, 'An instructional model for web-based e-learning education with a blended learning process approach', in *British Journal of Educational Technology*, 36(2), 217–235.

Bennett, S., Priest, A.-M. and Macpherson, C., 1999, 'Learning about online learning: an approach to staff development for university teachers', in *Australian Journal of Educational Technology*, 15(3), 207–221.

Blass, E. and Davis, A., 2003, 'Building on solid foundations: establishing criteria for e-learning development', in *Journal of Further and Higher Education*, 27(3), 227–245.

Biggs, J.B., 1999, 'What the student does: teaching for enhanced learning', in *Higher Education Research and Development*, 18(1), 55–76.

Boud, D., Keogh, R. and Walker, D., 1985, 'Promoting reflection in learning: a model' in D. Boud, R. Keogh and D. Walker (eds), *Reflection: Turning Experience into Learning*, London, Kogan Page, 18–40.

Cassidy, S., 2004, 'Learning style: an overview of theories, models and measures', in *Educational Psychology*, 24(4), 419–444.

Caudron, S., 1999, 'Free agent learner', in *Training and Development*, 53(8), 27–31.

Coles, C., 1997, 'Is problem-based learning the only way?' in D. Boud and G. Feletti (eds), *The Challenge of Problem-Based Learning*, 2nd edn, London, Kogan Page, 313–325.

Creswell, J.W., 1994, *Research Designs: Qualitative and Quantitative Approaches*, Thousand Oaks, CA, Sage.

DfES, 2003, *Towards a Unified e-Learning Strategy*, London, Department for Education and Skills.

DfES, 2005, *Harnessing Technology: Transforming Learning and Children's Services*, London, Department for Education and Skills.

Duggleby, J., Howard, J., Butler, K., Williams, L., Cooke, M., Cotton, C. and Schmoller, S., 2002, *Effective Online Tutoring Guidelines*, Sheffield College, on behalf of the South Yorkshire Further Education Consortium. http://www.jisc.ac.uk/techlearn_etutor.html (accessed 29 March 2005).

Ellis, R.A. and Calvo, R.A., 2004, 'Learning through discussions in blended environments', in *Educational Media International*, 41(3), 263–274.

Fallah, M.H. and Ubell, R., 2000, 'Blind scores in a graduate test: conventional compared with web-based outcomes', in *Asynchronous Learning Networks Magazine*, 4(2), htttp://www.sloan-c.org/publications/magazine/v4n2/fallah.asp

Fisher, J., 2004, 'The potential and use of electronic conferencing: a study of women's involvement in a global context', in *Information Development*, 20(4), 242–247.

Galagan, P.A., 2000, 'E-learning revolution', in *Training and Development*, 54(12), 25–30.

Geoghegan, W., 1995, 'Stuck at the barricades: can information technology really enter the mainstream of teaching and learning?', in *Change*, 27(2), 22–30.

HEFCE, 2005, *Strategy For e-Learning*, Bristol, Higher Education Funding Council for England.

Hostetter, C. and Busch, M., forthcoming, 'Measuring up online: the relationship between social presence and student learning satisfaction', in *Journal of the Scholarship of Teaching and Learning*.

Jones, P., Packham, G., Miller, C. and Jones, A., 2004, 'An initial evaluation of student withdrawals within an e-learning environment: the case of e-College Wales', in *Electronic Journal on e-Learning*, 2(1), 113–120.

Kolb, D.A., 1984, *Experiential Learning: Experience as the Source of Learning and Development*, Englewood Cliffs, NJ, Prentice-Hall.

Koos, W. and Betty, C., 2004, 'Learning productivity: a case analysis of the "e-BOSNO" course for manager teams', in *British Journal of Educational Technology*, 35(4), 443–460.

Laurillard, D., 2001, 'The e-university: what have we learned?', in *International Journal of Management Education*, 1(2), 3–7.

Laurillard, D., 2002, *Rethinking University Teaching: A Conversational Framework for the Effective Use of Learning Technologies*, 2nd edn, London, RoutledgeFalmer.

Leedy, P.D., 1997, *Practical Research: Planning and Design*, 6th edn, Englewood Cliffs, NJ, Prentice Hall.

Linser, R. and Ip, A., 2002, 'Beyond the current e-learning paradigm: applications of role play simulations (RPS) – case studies', in G. Richards (ed.) *E-Learn 2002 – Proceedings of the World Conference on E-Learning in Corporate, Government, Healthcare, and Higher Education, Chesapeake, VA*, Association for the Advancement of Computing in Education 606–611.

Littlejohn, A. and Higgison, C., 2003, *A Guide For Teachers: E-learning Series No.3*, York, Learning and Teaching Support Network.

MacKenzie, N. and Staley, A., 2000, 'Online professional development for academic staff: putting the curriculum first', in *Innovations in Education and Teaching International*, 38(1), 42–53.

Macpherson, A., Elliot, M., Harris, I. and Homan, G., 2004, 'E-learning: reflections and evaluation of corporate programmes', in *Human Resource Development International*, 7(3), 295–313.

Marton, F. and Säljö, R., 1976a, 'On qualitative differences in learning I: outcome and process', in *British Journal of Educational Psychology*, 46 (Part 1, February), 4–11.

Marton, F. and Säljö, R., 1976b, 'On qualitative differences in learning II: outcome as a function of the learner's conception of the task', in *British Journal of Educational Psychology*, 46 (Part 2, June), 115–127.

Moodle Sites, 2005, http://moodle.org/sites/ (accessed 29 March 2005).

Moore, G.A., 1991, *Crossing the Chasm: Marketing and Selling Technology Products to Mainstream Customers*, New York, Harper Collins.

Morón-Garcia, S., 2004, 'Understanding the approach to teaching adopted by users of virtual learning environments' in C. Rust (ed.), *11th Improving Student Learning Symposium: Theory Research and Scholarship*, Oxford, Oxford Centre for Staff Development, 235–248.

Murray, P.E., Donohoe, S. and Goodhew, S., 2004, 'Flexible learning in construction education: a building pathology case study', in *Structural Survey*, 22(5), 242–250.

NESTA Futurelab, 2004, *Literature Review in Games and Learning, Report 8*, Bristol, NESTA.

Palloff, R. and Pratt, K., 1999, *Building Learning Communities In Cyberspace: Effective Strategies for Online Classroom*, San Francisco, CA, Jossey-Bass.

Palloff, R. and Pratt, K., 2003, *The Virtual Student: A Profile and Guide to Working with Online Learners*, San Francisco, CA, Jossey-Bass.

Piccoli, G., Ahmad, R. and Ives, B., 2001, 'Web-based learning environments: A research framework and a preliminary assessment of effectiveness in basic IT skills training', in *MIS Quarterly*, 25(4), 401–426.

Race, P., 1998, *500 Tips for Open and Flexible Learning*, London, Kogan Page.

Race, P., 2001, *The Lecturers Toolkit*, 2nd edn, London, Kogan Page.

Russell, T., 1999, *The No Significant Difference Phenomenon*, Montgomery, AL, International Distance Education Certification Center.

Salmon, G., 2000, *E-moderating: The Key to Teaching and Learning Online*, London, Kogan Page.

Sandelands, E. and Wills, M., 1996, 'Creating virtual support for lifelong learning', in *The Learning Organisation*, 3(5), 26–31.

Skinner, E., 2004, 'Engaging students in active online participation' in M. Healey and J. Roberts (eds), *Engaging Students in Active Learning: Case Studies in Geography, Environment and Related Disciplines*, Cheltenham, Geography Discipline Network 68–74.

SLICE, 2002, *Handbook on Student-centred Learning in Construction Education*, Plymouth, University of Plymouth.

Staley, A. and MacKenzie, N. (eds), 2001, *Computer Supported Experiential Learning*, UCE Birmingham, Learning Methods Unit.

Thorne, K., 2003, *Blended Learning: How to Integrate Online and Traditional Learning*, New Jersey, Kogan Page.

Webb, E., Jones, A., Barker, P. and Schaik, P., 2004, 'Using e-learning dialogues in higher education', in *Innovations in Education and Teaching International*, 41(1), 93–104.

Wiles, K. and Littlejohn, A., 2003, 'Supporting sustainable e-learning: a UK national forum', in G. Crisp *et al* (eds), *Interact, Integrate, Impact: Proceedings of the 20th Annual Conference of the Australasian Society for Computers in Learning in Tertiary Education, Adelaide, December 2003*, Adelaide, ASCILITE, 730–734.

Wilson, G. and Stacey, E., 2004, 'Online interaction impacts on learning: teaching the teachers to teach online', in *Australasian Journal of Educational Technology*, 20(1), 33–48.

Yin, R.K., 2003, *Case Study Research: Design and Methods*, 3rd edn, London, Sage.

ARTICLE

An Ontology of Construction Education for E-learning via the Semantic Web

Vian Ahmed, Azmath Shaik and Ghassan Aouad

Abstract

Recent developments in technologies make the World Wide Web more intelligent and provide higher-level services to its users through the Semantic Web. The Semantic Web is an extension of the current World Wide Web, promoting information that is intended not only for human readers, but also can be processed by machines, enabling intelligent information services and semantically empowered search engines. This has a number of important implications for Web-based education, since Web-based education has become an important branch of educational technology. Although this technology is relatively new with limited use within various educational domains, its potential has not yet been explored within the construction education domain. The integration of the Semantic Web will provide platform independence and intelligence in Web-based educational applications for construction education and a solution to e-learning tools that have become redundant as a result of the fast developments in technologies. This paper gives an overview of the development of e-learning in construction, highlighting the key pedagogical and technical concepts and enablers to Semantic Web technology. The paper also describes an approach adopted for the development of 'ontology' within the construction domain and its integration to define the educational content, both semantically and pedagogically to enable a platform-independent architecture for both learners and educators.

■ *Keywords* – Construction education; ontology; semantic web; e-learning

INTRODUCTION AND BACKGROUND

The instruction methods used in the majority of construction, engineering and management curricula rely on traditional methods such as exposing students to applied science courses (Moonseo *et al*, 2003). These traditional teaching methods, however, are often not fully capable of providing students with all the skills necessary to solve the real world problems encountered in construction or conveying complex engineering knowledge effectively. Also, curricula often convey knowledge in fragments in a series of courses.

Ideally, visits to construction sites or site training would constantly complement the more conventional classroom instructional tools. However, there are various complicating issues that make it impossible to rely on the sites. Foremost, the instructor cannot control the availability of a project at the necessary stage of completion. Also, visits of larger groups to construction sites may not be welcome, involve risk and are unpractical. Finally, the high cost of site training is a further impediment to its extensive use for construction education. Therefore, the use of information technology for educational purposes, such as e-learning in general and simulations in particular, has the potential and can act as an excellent tool to complement construction and management education. This section describes the state of the art of e-learning in construction and the related pedagogical concepts that can be adopted to enable effective learning in construction.

E-LEARNING IN CONSTRUCTION

The implementation of e-learning into construction education had a slow start but is now catching up with other fields. During the 1990s, a number of UK initiatives were launched to support the development and integration of e-learning within higher education. This process can be broken down into three phases. Phase 1 (1992–1996) formed the era of stand-alone applications; phase 2 (1996–1998) was the generic 'online' application phase; and phase 3 (1998–present) moved towards a Web-based and wireless mobile learning phase. At present, there is a major focus on Web-based learning (e-learning) which is slowly and steadily moving towards mobile learning (m-learning) (Chris, 2003), even though the cost of its implementation is a bit of a hindrance (John, 2003).

One of the initiatives is the Teaching and Learning Technology Program (TLTP) fund which supported the Visual CAL (Computer Aided Learning) Project in 1999, which produced a database of shared images described using metadata. The Joint Information System Committee (JISC) also funded a number of similar initiatives to look into e-learning standards to bridge content and systems. Since then, the development of e-learning tools for construction education has been continuously growing. Their development, however, is costly and time consuming and they soon become out of date. This is mainly because of fast changes in the technology and their incompatibility with new systems or platforms and as a result of instructional software available being locally effective, but globally fragmentary (Molyneux, 2002).

Educators must learn to teach in the context of new pedagogical models and to use online tools, so that they can perform the core of their job more effectively, efficiently and with more appreciation (Koper, 2004). Educational systems must be flexible in that they must be easy to adapt to new and changing user requirements. The state-of-the-art way of dealing with open requirements is to build systems out of reusable components to a plug-in architecture (EL Saddik and Fischer, 2000). The functionality of such systems can be changed or extended by substituting or plugging in new components. Although the component-based solutions developed to date are useful, they are inadequate for those building component-based interactive learning environments in which the components must respond to the meaning of the content as well as its form and representation. Construction educators and technologists must take the lead in promoting computer literacy in their curriculum and continue to develop new courses, delivery styles and software applications through continued research activities (Charles *et al*, 2004).

On the other hand, educational software has been developed with the potential for online learning as complete packages, commercially prepared and disseminated. However, they lack cohesion as an organized collection because every software application is vertically engineered to comply within its specific domain. This makes their reusability and sharing more difficult and can lead to maintenance and deployment difficulties as restrictive platform requirements accumulate over time. One of the main highlights of this paper is to try to solve this problem by adopting standards.

E-LEARNING PEDAGOGY

To develop effective e-learning components, it is important to develop an understanding of the pedagogical concepts that enable a level of intelligence within the e-learning content. Pedagogy is defined as the means of transmission of professional knowledge, skills and culture to the learners and affects how learners retain the knowledge and culture they need.

According to Bloom's cognitive taxonomies, learning begins with the reception of information (Anderson *et al*, 2000). The instructor's ability to provide a learning environment that will engage the students' interest is crucial to this process. In addition, a student's interest, readiness and willingness to engage in the learning process are also important. Another key factor is the students' learning style. A student's preferred and unique learning style is still a disputed fact about how students (at any level) learn (Grasha, 1996). There is a learning style variation among students. In some respect, the variation in learning style is desirable because it leads to different types of innovation and develops a wide variety of talents. However, from an educational perspective, favouring one learning mode over another potentially results in a mismatch between the learning styles of the learner and educator (Ahmed, 2000). This mismatch or disparity is frustrating to the student whose learning style is not compatible with that of the educator. This is termed in the literature as educating 'around the cycle' (Kolb, 1984).

Kolb's model considers the cognitive processes preferred by students in acquiring and understanding information. Felder (1996) gives examples of how educators used learning style models to successfully educate. The learning style models mentioned have been the basis for developing psychometric tools for evaluating the learning style of students and instructors. The learning style tools have been developed to assess learning styles and design instruction accordingly. Efforts have been made to understand students' preferential style of learning and to design course instruction to 'cater' for all learning modes. Different learning styles can be defined as (Felder and Soloman, 2001):

- Processing (active/reflective). Active learners learn best by doing something physical with the information, whereas reflective learners do the processing in their heads.
- Perception (sensitive/intuitive). Sensitive learners prefer data and facts, whereas intuitive learners prefer theories and interpretations of factual information.
- Input (visual/verbal). Visual learners prefer charts, diagrams and pictures, whereas verbal learners prefer the spoken word.
- Understanding (sequential/global). Sequential learners enjoy making connections between individual steps, whereas global learners need to form the 'big picture' before the individual pieces are put together.

There is, therefore, a need to harmonize the educating and learning material to different styles of students' minds and learning styles. It is recognized that learning is promoted when information resources and ideas are readily available to educators and learners. Ready access to learning resources, such as knowledge, best practices, problem/solution cases is indeed a key in order to support the high standards set by the employment market. The Index of Learning Styles (ILS) model (Felder and Soloman, 2001) can be used as a framework to determine the learning style preference of educators and their learners at both the undergraduate and graduate levels (construction education domain). The ILS model is based on the Felder-Silverman Learning Style Model (Felder, 1996) and is comprised of 44 questions with just two options to tick for each. The combinations of answers determine the respondent's learning style preference on four different dimensions. The learning objects are assigned to these dimensions satisfying the needs of the learners.

These pedagogical approaches are not new, but their implementation is not widely spread. Their implementation is on a small scale in construction education. This paper considers the pedagogical aspects of e-learning as a key factor in the development of e-learning content. Such content is referred to as learning objects.

LEARNING OBJECTS AND CONTENT

A review of relevant literature reveals a number of definitions of the term learning object (LO) proposed by different authors. These definitions are often used interchangeably and can be contradictory or confusing. This, however, is understandable, due to the recent developments and acceptance of the field of e-learning. This section highlights some of these definitions.

Mohan and Brooks (2003) define an LO as 'a digital learning resource that facilitates a single learning objective and which may be reused in a different context'.

South and Monson (2003) use the term 'media object'. They define such an object as 'digital media that is designed and/or used for instructional purposes. Such objects range from maps and charts to video demonstrations and interactive simulations.'

Wiley (2003) defines LOs as elements of a new type of computer-based instruction grounded in the object-oriented paradigm of computer science. They allow instructional designers to build small (relative to the size of an entire course) instructional components that can be reused a number of times in different learning contexts. They are generally understood to be digital entities deliverable over the Internet, meaning that any number of people can access and use them simultaneously.

LEARNING CONTENT MODELS

According to Verbert and Duval (2004), learning content can be categorized into content fragments, content objects and learning objects:

- Content fragments are learning content elements in their most basic form, such as text, audio and

TABLE 1 Comparison of terms used in different LCMs

MODEL	CONTENT FRAGMENTS	CONTENT OBJECTS	LEARNING OBJECT	OTHER TERMINOLOGY USED		
Learnativity	Raw media	Information object	Learning object	Aggregate assemblies		Collections
SCORM	Assets	–	SCO	Content aggregation		–
CISCO	Content items	RIO	RLO	–	–	–
Netg	–	–	Topic	Lesson	Unit	Course

SCORM – Sharable content object reference model

SCO – Sharable content object

RIO – Reusable information object

RLO – Reusable learning object

video, and can be time-based (audio, video and animation) and static (photo, text, etc.).

- Content objects are sets of content fragments. They aggregate content fragments and add navigation.
- Learning objects consist of content objects and a learning objective. They define a logical relationship between their components.

The terms content fragments, content objects and learning objects are used in the development of the ontology (term as discussed in 'Semantic Web and Ontology' below). Ontology provides guidance to describe learning objects and their components in a consistent fashion, facilitating sharing and reuse of both learning objects and their components. This strategy can be and is adopted to get an efficient ontology in terms of learning content.

A review of literature concerning the different learning content models (LCM) made available, shows that they often differed in usage of terms for each category of learning content as defined in Table 1 (Verbert and Duval, 2004).

Thus in order to utilize these learning objects efficiently, in terms of data exchange, reusability or interoperability, there is a need for the standardization of these learning objects both technically and also pedagogically.

STANDARDIZATION OF LEARNING CONTENT

Standardization is considered the ultimate solution to the problem of data exchange. In order to reuse content from one system to another, it is important for learning objects to be standardized. A number of global efforts are being made to develop standards, specifications and reference models for both learning technologies and Web-based technologies. These standards deal with interoperability among different learning management systems (LMSs) and different types of content in a LMS or Web-based learning environment. These standards will enable pedagogical content to be shared and reused by all sorts of e-learning systems.

The concept of Semantic Web and the development of ontologies will aid in fulfilling these needs with respect to Web-based learning and are explained in detail in the following sections.

SEMANTIC WEB AND ONTOLOGY

The Semantic Web is an extension of the current World Wide Web. The next generation of the Web is often characterized as the Semantic Web where information will no longer only be intended for human readers, but also for processing by machines, enabling intelligent information services and semantically empowered search engines. The hypertext pages that present information to humans remain, but a new layer of machine understandable data is added to allow computers to participate on the Web in new ways and, if possible, using ontology (Berners-Lee *et al*, 2001).

Ontology defines the terms used to describe and represent an area of knowledge. Ontologies are used by people, databases and applications that need to share domain information (a domain is just a specific subject area or area of knowledge, like medicine, tool manufacturing, real estate, automobile repair, financial management, etc.). Ontologies include computer-usable definitions of basic concepts in the domain and the relationships among them. They encode knowledge within a domain and also knowledge that spans domains. In this way, they make knowledge reusable. Ontologies make the learning objects or the e-resources more intelligent than what the metadata does, as the ontology includes the logic-based description of the learning objects.

Most academics are not ontological engineers, domain experts or logicians, so it is unlikely that they will be able to read, sort through and grasp how to apply large ontologies, much less construct their own. In order to make the development of ontologies easy and robust, there are tools available for these purposes known as ontological editors (Denny, 2004).

The Semantic Web requires interoperability on the semantic level. Semantic interoperability requires standards not only for the syntactic form of documents, but also for the semantic content. Proposals aiming at semantic interoperability are the results of recent W3C standardization efforts, notably eXtensible Markup Language (XML); XML Schema; Resource Description Framework (RDF); RDF Schema and the final W3C recommendation for the Semantic Web, i.e. the Web Ontology Language (OWL, 2004) used to represent the ontologies.

The Educational Semantic Web is based on three fundamental affordances (Anderson and Whitelock, 2004), namely, the capacity of:

- effective information storage and retrieval
- non-human agents to augment the learning and information retrieval
- Internet to support, extend and expand communications capabilities of humans in multiple formats across the bounds of time and space.

The generic Semantic Web model is shown in Figure 1 (Verbert and Duval, 2004) whereby the learning content is developed by the educator along with its pedagogical value and submitted in a repository based on semantic technology consisting of its respective domain ontology. The learning content can then be searched, retrieved and reused with the help of semantic agents, including pedagogical agents, based on the ontology which gives a rich and logical meaning to the content.

Ontologies are frequently used in e-learning platforms, first, to organize their content and, second, to facilitate retrieval. The learning object metadata

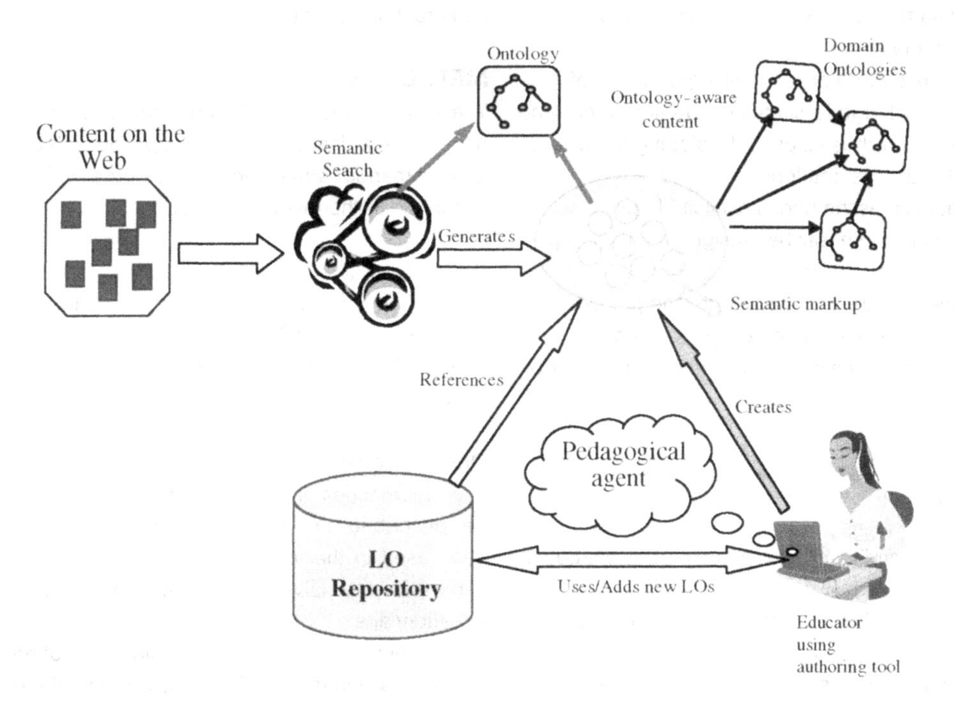

FIGURE 1 Generic semantic web model (Verbert and Duval, 2004)

(LOM, 2003) has been introduced to enable sharing and reuse of learning objects across platforms. However, this standard only offers a limited number of attributes to describe pedagogical aspects of a learning object (such as the interactivity type, the kind of learner it is suitable for, etc.). Therefore, the LOM standard is not suitable to fully capture pedagogical knowledge (Allert *et al*, 2002; Pawlowski, 2002).

ONTOLOGY DEVELOPMENT

The reasons for ontology development are to:

- 'Share common understanding of the structure of information among people or software agents.' It is the most common goal in developing an ontology. If websites share and publish the same underlying ontology of terms they all use, then computer agents can extract and aggregate information from these different sites.
- 'Analyse domain knowledge.' Analysing domain knowledge is possible once a declarative specification of the terms is available. Formal analysis of terms is extremely valuable when both attempting to reuse existing ontologies and extending them.
- 'Enable reuse of domain knowledge.' Reuse of domain knowledge facilitates integration of existing ontologies and thus helps in developing a precise, detailed and large ontology.
- 'Make domain assumptions explicit.' Explicit assumptions can easily be changed if knowledge about the domain changes.
- 'Separate domain knowledge from the practical knowledge.' Develop an ontology based on components and characteristics first and then apply practically.

ONTOLOGY HIERARCHY

An ontology has the following hierarchical structure:

- *packages* (broad categories) which group various ...
- *classes* (items) which have various ...
- *restrictions* (attributes or properties) which describe the relationship between classes
- *individuals* (defined sets of values a class is allowed to have)
- *sub-classes*
- *links to external resources* (databases, ontologies).

Ontology is considered as an integration of various metadata standards, namely, the Learning Object Metadata (LOM) Standard, Dublin Core Metadata Initiative (DCMI), Digital Object Identifier (DOI) and the bodies that developed these standards are adopting the ontology concept and are moving to its common implementation (Figure 2) (Dekkers, 2004).

The present developments in educational technologies to deliver knowledge leads to the implementation of these technologies in every domain of knowledge. It has been found that there is a genuine need for adopting the Semantic Web technology standards, which have been projected as the main medium to encourage the interoperability of educational resources and data in general.

The rest of the paper describes the methodological approach for the development of an online learning content repository for construction education based on the concept of the Semantic Web and the implementation of an ontology related to the domain of construction education.

METHODOLOGY

The project was divided into three phases: ontology development, learning content repository development and semantic application development. These are described in the following sections.

PHASE 1: ONTOLOGY DEVELOPMENT

The subject of study of this phase is the design and development of ontology for the construction education domain. The stages of the methodology are described below.

Ontology design

The approaches in the design of an ontology can be described in Table 2. These approaches have been used in the form of discussion with the domain experts, extensive literature research and a questionnaire.

A general criterion for the development of education ontology is illustrated in Figure 3, whereby the ontology for the domain is based on user needs and the

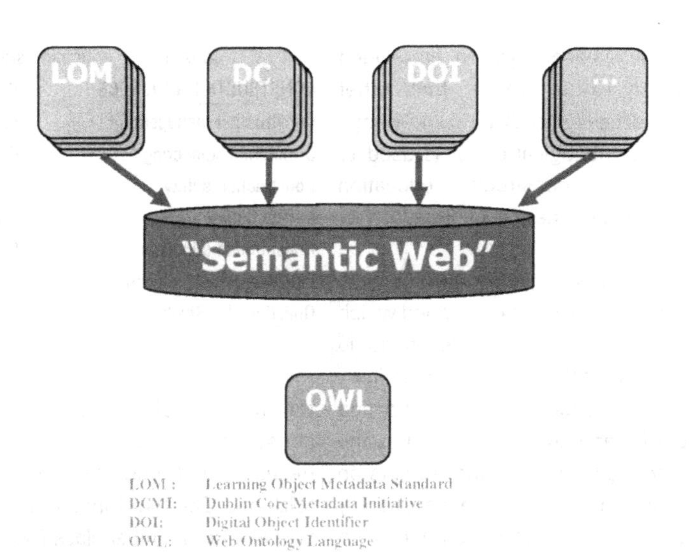

FIGURE 2 Ontology: integration of various metadata standards

TABLE 2 Ontology design approaches

APPROACH	BASIS FOR DESIGN
Inspiration	Individual viewpoint about the domain
Induction	Specific case within the domain
Deduction	General principles about the domain
Synthesis	Set of existing ontologies
Collaboration	Multiple individuals' viewpoints about the domain, possibly coupled with an initial ontology as an anchor

pedagogy involved and thus describes the content in terms of these needs and context.

Ontology editor

An ontology editor is a tool to design and develop an ontology for a domain having a user-friendly interface. There are numerous free and open-source ontology engineering tools on the Internet used for the development of ontologies. More information about these and other tools can be found in a survey by Denny (2004). The most popular ontology editor being used is the Protégé toolkit (Protégé, 2004) as it has adopted the recent recommendation of the W3C, i.e. OWL (Web Ontology Language) standard and is chosen to develop and maintain the ontology for our domain of knowledge – construction education.

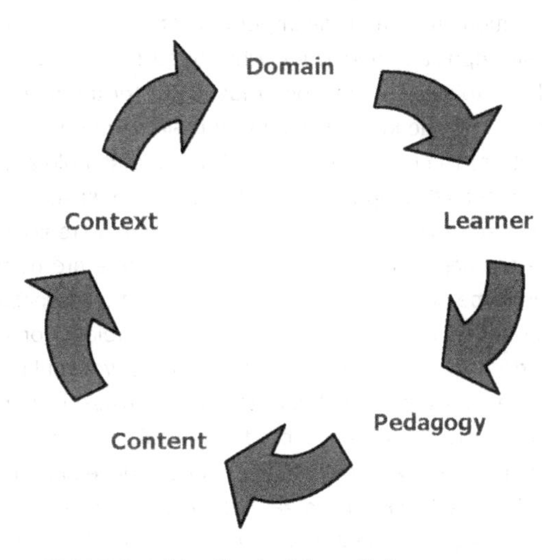

FIGURE 3 Educational ontology criteria

Protégé is an ontology editor and a knowledge-based editor. Protégé is an open-source Java tool that provides an extensible architecture for the creation of customized knowledge-based ontologies. It acts as a *tool* that allows the user to construct a domain ontology, customize data entry forms and enter data. It also acts as a *platform* that can be extended with graphical widgets for tables, diagrams and animation components to access other knowledge-based system embedded applications and, finally, as a *library* that other applications can use to access and display knowledge bases. Thus, the Protégé ontological editor is used to develop the ontology for the construction education domain described in detail in the next section.

Ontology for construction learning objects

A sample and initial ontology has been developed which included a set of sub-domains (further described in detail) and their inter-relationships. After preliminary stage of studying the various educational standards and Web standards (XML, RDF and OWL), which act as the main modes of delivery of e-learning to the learners on the Web, a related set of standards were chosen, mainly the metadata and Semantic Web standards. A preliminary pilot questionnaire was conducted to develop the initial ontology. The purpose of the questionnaire was to help ascertain what kinds of information are potentially important for a construction domain repository and study what need is catered for by the ontology. The ontology must cater for the needs of the construction education domain learners and educators. The initial focus is to develop a sample and experimental description of the domain – the terms that experts see as important data entries and that are likely to be the areas of maximum interest. If there are new terms to be added, it can be decided whether to pass these and enter them into the sample ontology, or if they do not 'fit', make our own controlled vocabulary. The questionnaire helped to obtain a clear classification of the domain in order to produce ontologies for the domain. An online questionnaire is under development to collect feedback for the sample ontology developed so far and get thorough input from a group of domain experts of construction education.

After a preliminary survey (questionnaire) of the construction education domain, an initial ontology was developed that would cater for the needs of adopting the Semantic Web standard for the educational server, which helps in the description and delivery of construction education resources on the Web.

The construction education domain has been classified into sub-domains by studying the output of an initial pilot questionnaire developed for the purpose. A group of domain experts was involved and a study of the major global bodies relating to building and construction, namely, the International Council for Research and Innovation in Building and Construction (CIB) and Chartered Institute of Building (CIOB) helped to make up an initial classification of domain. The sub-domains are listed in Table 3 and shown in Figures 6 and 7 as sub-classes from a screenshot of Protégé ontological editor.

TABLE 3 Construction education sub-domains

CONSTRUCTION GRAPHICS	PROJECT MANAGEMENT
Construction management	Facilities management
Structural engineering	Construction law
Construction safety	Construction design
Design theory	Quantity surveying
Construction surveying	Cost estimating
Planning and scheduling	Construction accounting
Questions for Shaik	–

The instances applicable to various sub-classes, along with the levels of education and the related pedagogy classes, are illustrated diagrammatically in Figures 4 and 5. The pedagogical aspect is an important ingredient of education. It comprises various instructional theories and their related styles, and various learning theories and their related styles. Keeping in mind the importance of pedagogy, related learning and instructional styles for the domain have been incorporated and linked to various sub-domains.

Ontological diagrams (Figures 5, 7, 8 and 9) – using a plug-in, OWLViz, in Protégé – give a good view and a clear visual understanding of the whole knowledge domain and the relations between various sub-domains and their pedagogical link.

A representation of learning object types related to construction education and the categories is shown in Figure 8.

An example of the ontology developed for a sub-domain (project management) is shown in Figure 9.

Figure 10 shows the metrics (statistics) of the sample ontology developed illustrating its 55 classes, 77 slots (properties), 10 facets and 44 instances (terms mentioned in 'Ontology Hierarchy').

The Protégé ontological editor has been found to be the best among various editors available in terms of ease of use, implementation of the recently released

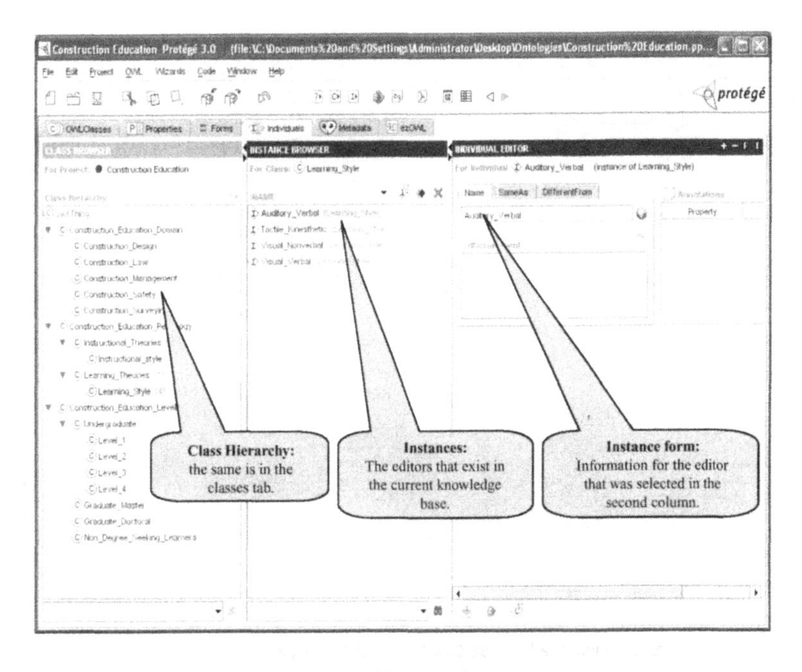

FIGURE 4 Protégé interface (classes and instances)

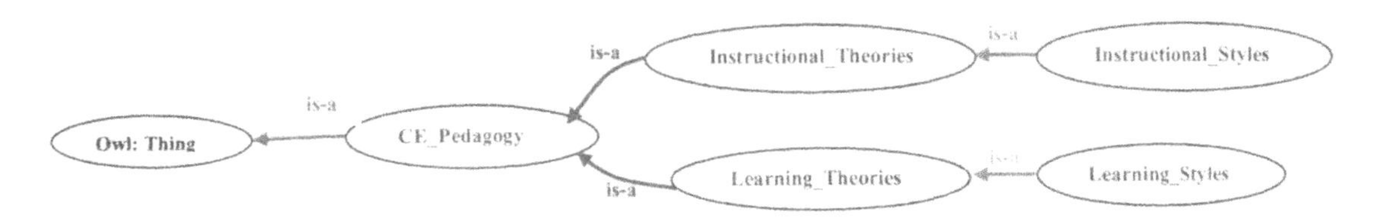

FIGURE 5 OWLViz plug-in representation

OWL standard (February 2004) and the various plug-ins available like the OWLViz tool. The ontologies thus developed enable learners to get the most suitable information retrieval for a search using a semantic search engine incorporated into the Semantic Web e-learning portal interface running on the construction education semantic server described below.

PHASE II: LEARNING CONTENT REPOSITORY

This phase aims at the simultaneous development of the learning content repository along with the ontology development, which is used to describe the learning content related to the construction education domain. A construction domain educational server (Figure 11) with semantic content and semantic profiles for the learning content will be developed in this phase along with the process of refining ontology development, described above, with regard to the pedagogical needs and benefits and design of ontology development.

PHASE III: APPLICATION DEVELOPMENT

This phase involves the development of an e-learning application framework using a Semantic Web application development toolkit recommended by the W3C consortium. It could be used to develop a learning content repository for construction education using the ontology for construction education describing all the

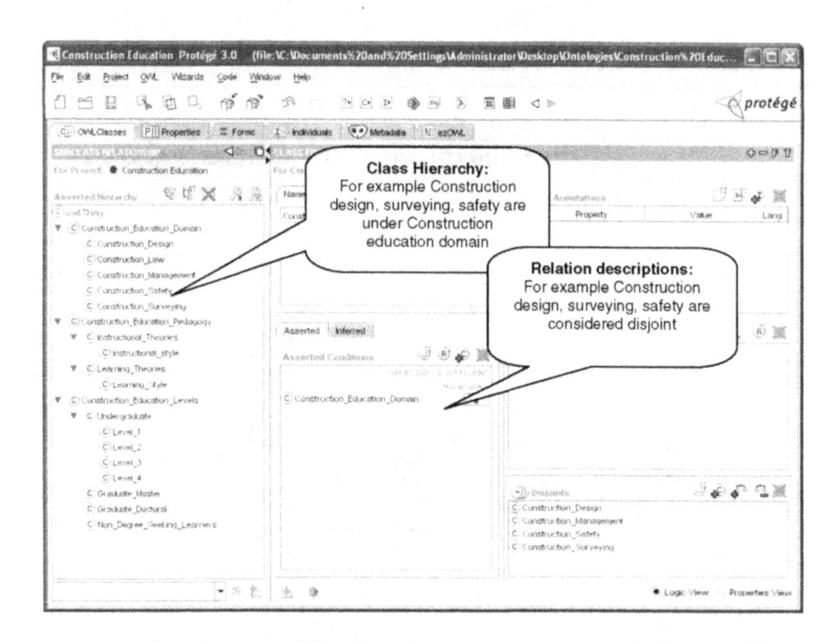

FIGURE 6 Protégé interface (classes and logical relationships)

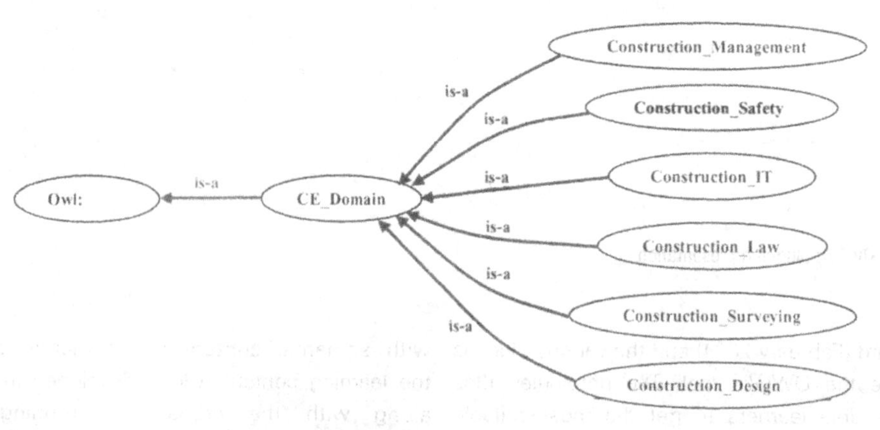

FIGURE 7 OWLViz representation of the domain

learning content both logically and pedagogically. In fact, it will be a repository of the learning content, such as websites, word documents, pdf files, PowerPoint presentations, simulations, etc., plus logical descriptions about these objects. A most important aspect of the application is the ontology. The addition of such semantic information to the learning content will be done via OWL, using dynamically generated Web pages, personalized for every user. Since the information is stored in the form of machine-understandable OWL statements, it can be used by the application agents. A domain-specific ontology has been developed in order to give meaning to these OWL statements and can be used within the application to give logical meaning and relationship to the learning content.

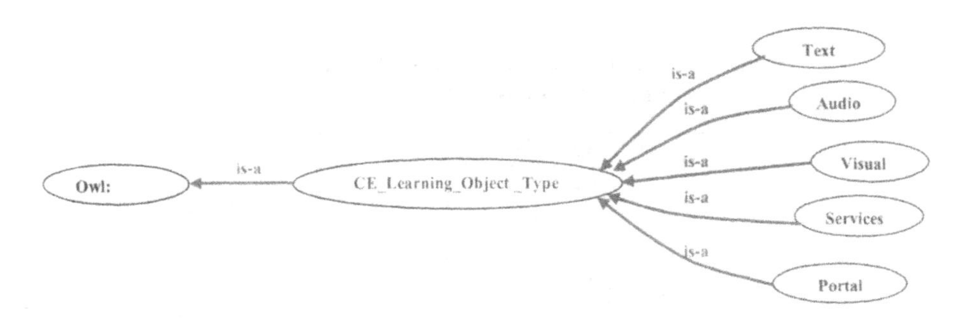

FIGURE 8 OWLViz representation of types of learning objects

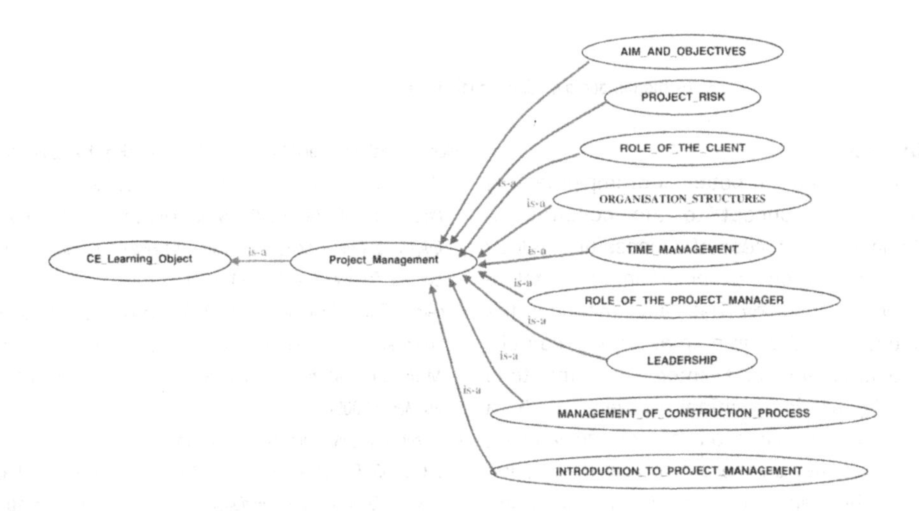

FIGURE 9 OWLViz representation of sub-domain project management

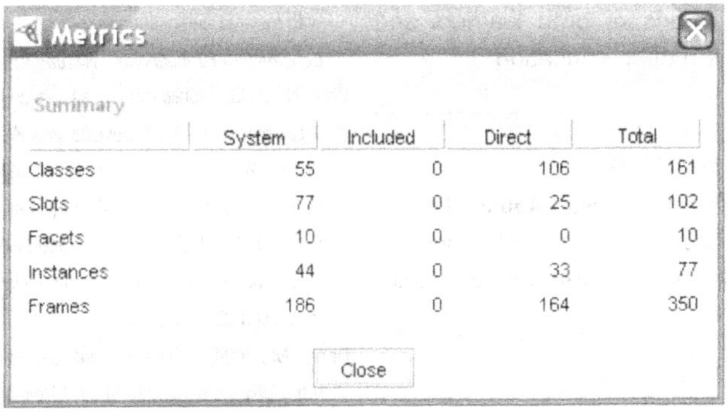

Metrics

Summary

	System	Included	Direct	Total
Classes	55	0	106	161
Slots	77	0	25	102
Facets	10	0	0	10
Instances	44	0	33	77
Frames	186	0	164	350

Close

FIGURE 10 Metrics (statistics) of the ontology developed

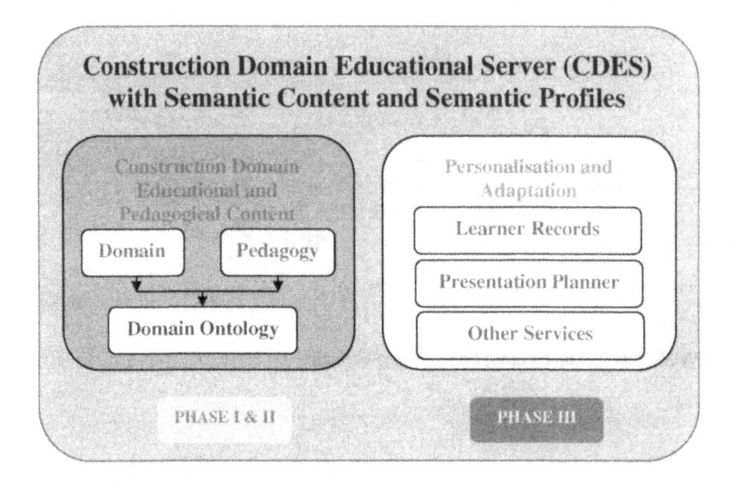

FIGURE 11 Construction domain educational server

CONCLUSIONS

This paper has explored the potential of implementing the Semantic Web concept to the construction education domain. The process of successful ontology development is a step forward for the implementation of Semantic Web technology. The paper mentions the various benefits of this implementation, namely, reusability, adaptability and interoperability thus enabling intelligent information retrieval with a semantically powered search engine for the learning object repository based on the ontological description of the domain. This paper describes the research and development carried out for a project involving the implementation of the Semantic Web concept to the construction education domain in the form of an e-learning semantic portal environment to incorporate the various pedagogical needs of both learners and educators involved in construction education.

AUTHOR CONTACT DETAILS

Azmath Shaik, Vian Ahmed and Ghassan Aouad: School of Construction and Property Management, University of Salford, Salford, Lancashire, M5 4WT, UK, e-mail: a.k.shaik@pgr.salford.ac.uk

REFERENCES

Ahmed, V., 2000, 'The effectiveness of computer aided learning in construction', PhD Thesis, Loughborough University.

Allert, H., Richter, C. and Nejdl, W., 2002, 'Learning objects and the Semantic Web. Explicitly modelling instructional theories and paradigms', *Proceedings of E-Learn 2002: World Conference on E-Learning in Corporate, Government, Healthcare, and Higher Education, Montreal, Canada, October 15–19*, 1124–1127.

Anderson, T. and Whitelock, D., 2004, 'The educational semantic web: visioning and practising the future of education', in *Journal of Interactive Media in Education*, 1–17, http://www-jime.open.ac.uk/2004/1 (accessed 25 March 2005).

Anderson, L.W., Wittrock, M.C., Krathwohl, D.R., Mayer, R.E. and Pintrich, P.R., 2000, *Taxonomy for Learning, Teaching and Assessing: A Revision of Bloom's Taxonomy of Educational Objectives*, Abridged edn, 2001, Pearson Education.

Berners-Lee, T., Hendler, J. and Lassila, O., 2001, 'The Semantic Web', in *Scientific American*, 284(5), 34–43.

Charles, W.B., Bruce, F., Tim, W. and Michael, D.N., 2004, 'Distance education with Internet audio/video technology', in *The International Journal of Construction Education and Research*, 1(1), 45–60.

Chris, N., 2003, 'Mobile learning as service offering with near-term technologies', in *Learning with Mobile Devices, 2nd World Conference on MLearning, MLearn 2003*, 117–126, http://www.m-learning.org/docs/MLEARN_2003_Book_of_Abstracts_May_03.pdf (accessed 25 March 2005).

Dekkers, M., 2004, *DCMI Status Report March-September 2004*, http://dublincore.org/news/communications/statusreport-200409.shtml (accessed 25 Mar 2005).

Denny, M., 2004, *A Survey of Ontology Editors by Michael Denny*, http://www.xml.com/2004/07/14/examples/Ontology_Editor_Survey_2004_Table_Michael_Denny.pdf (accessed 25 March 2005).

ElSaddik, A. and Fischer, S., 2000, 'ITBeanKit: An educational middleware framework for bridging software technology and education', in *World Conference Educational Multimedia, Hypermedia and Telecommunications*, (1), 1323–1324.

Felder, R.M., 1996, 'Reaching the second tier: learning and teaching styles in college science education', in *Journal of College Science Teaching*, 23(5), 286–290.

Felder, R.M. and Soloman, B.A., 2001, *Learning Styles and Strategies*, http://www.ncsu.edu/effective_teaching/ILSdir/styles.htm, North Carolina State University (accessed 25 March 2005).

Grasha, A., 1996, *Teaching with Style: A Practical Guide to Enhancing Learning by Understanding Teaching and Learning Styles*, Pittsburgh, Alliance Publishers.

John, T., 2003, 'Mobile learning – evaluating the effectiveness and the cost', in *Learning with Mobile Devices, 2nd World Conference on MLearning, MLearn 2003*, http://www.m-learning.org/docs/MLEARN_2003_Book_of_Abstracts_May_03.pdf (accessed 25 March 2005).

Kolb, D.A., 1984, *Experiential Learning: Experience as the Source of Learning and Development*, Englewood Cliffs, NJ, Prentice-Hall.

Koper, R., 2004, 'Use of the semantic web to solve some basic problems in education', in Journal of Interactive Media in Education, (6), http://www-jime.open.ac.uk/2004/6 (accessed 25 March 2005).

Mohan, P. and Brooks, C., 2003, 'Learning objects on the semantic web,' *International Conference on Advanced Learning Technologies, Athens, July 2003.*

Molyneux, S., 2002, 'The future of e-learning', in *Learning Lab Journal*, 1(3), 4–5, http://www.learninglab.org.uk/Journals/Learning LabJournal_vol1_issue3.pdf (accessed 25 March 2005).

Moonseo, P., Swee, L.C. and Yashada, I.V., 2003, 'Three success factors for simulation based construction education', in *Journal of Construction Education*, 8(2), 101–114.

OWL, 2004, *Web Ontology Language*, http://www.w3.org/TR/owl-ref (accessed 25 March 2005).

Pawlowski, J., 2002, 'Reusable models of pedagogical concepts – a framework for pedagogical and content design', *World Conference on Educational Multimedia, Hypermedia and Telecommunications*, (1), 1563–1568.

Protégé, 2004, 'Protégé Toolkit', http://protege.stanford.edu (accessed 25 March 2005).

South, J.B. and Monson, D.W., 2003, 'A university-wide system for creating, capturing, and delivering learning objects: the instructional use of learning objects', http://www.reusability.org/read/chapters/south.doc (accessed 25 March 2005).

TLTP Phase 3 Project, 1999, 'Implementation of computer imagery and visualization in teaching, learning and assessment' [online] available at http://www.le.ac.uk/tltp/contacts.html. [accessed 25 Mar 2005].

Verbert, K. and Duval, E., 2004, 'Towards a global architecture for learning objects: a comparative analysis of learning object content models', in *World Conference on Educational Multimedia, Hypermedia and Telecommunications*, (1), 202–208.

Wiley, D.A., 2003, 'Connecting learning objects to instructional design theory: a definition, a metaphor, and a taxonomy: the instructional use of learning objects', http://reusability.org/read/chapters/wiley.doc (accessed 25 March 2005).

ARTICLE

Learning to be Real Engineers

The Dam Game Simulation

Susan J. Gribble, David Scott, Mick Mawdesley and Saad Al-Jibouri

Abstract

Most engineering programmes around the world now ensure students develop specified graduate attributes and achieve clearly stated learning outcomes. Not only do engineering graduates require technical knowledge and skills, they also need to demonstrate that they have acquired competencies related to the more social aspects of engineering practice. Working in teams, communicating with people from diverse backgrounds and conducting themselves in an ethical and responsible way are some of these types of learning outcomes that are expected of engineering graduates. An 18-month study has been conducted with more than 250 undergraduate students at Curtin University of Technology (Perth, Australia) into the effectiveness of a simulation in developing these outcomes. In the study, close attention was paid to learning theory and research methodology associated with investigating educational settings. The information gathered focused on how students reacted to the simulation as a learning tool, the ways in which students used the simulation to learn and the learning outcomes students achieved through their learning experiences. The study demonstrated, in the main, that students believed that the simulation was an effective learning tool for them and they recognized that the simulation helped them to develop skills in applying their fundamental engineering knowledge to a civil engineering construction project. They also developed understanding about how their engineering decisions affected the workplace, people and the environment. Students were confident that the simulation taught them much about working as part of a professional team because they had to cooperatively plan, monitor, control and report on their project. Furthermore, the study showed that the simulation should be part of a holistic teaching and learning experience in which explicit teaching strategies are required so that students gain optimum learning by using the simulation.

■ *Keywords* – Civil engineering; simulation; constructivist learning; interpretive research

INTRODUCTION

Gilgeous and D'Cruz (1996) described the use of simulations in business management stretching back over many years and the use of management simulations for teaching and learning about project planning and control has long been documented in the literature (Scott and Cullingford, 1973). Furthermore, Au and Parti (1969) explored the use of a simulation using a civil engineering project with a significant amount of earthmoving as a basis. But not all games have to be complex and computer based, as Tommelein *et al* (1999) explained. Simulations can be run either manually or on a computer to illustrate the interaction of various personnel involved in a project. Indeed, the dam game simulation, central to this study has at various stages of its development relied on a great deal of manual calculation of results. More recently, the Internet has featured a variety of simulations as part of the learning environment in engineering education (Sawhney *et al*, 2001). For example, the idea of using computerized interfaces in manufacturing industries such as chemical processing and '[t]he trend employing computers to simulate, monitor and control manufacturing operations suggests an opportunity to modernize the pedagogy of engineering laboratories' (Wiesner and Lan, 2004: 195). More to the point:

> *Students are always excited by doing 'real' engineering. Simulations that engineers use to design everything from bridges to circuits are a good way to capitalize on this enthusiasm. They allow students to complete realistic designs in a reasonable amount of time, and there are versions for all engineering disciplines. (Wankat and Oreovicz, 2004: 45)*

Despite these developments, simulations are underused in engineering education because their functionality has not kept pace with the development of computer technology over the past decade. Most of them are still text based, while most 'work' computers have moved to a graphical user interface (GUI).

The dam game simulation combines the experience of students and classroom educators (both teachers and engineers) to provide a user interface and style of learning that is both interesting and informative. The instructional strategies of the simulation include:

- a realistic model of a construction project that will react in physical and financial terms to the decisions made and actions taken by the students
- reports as might be expected on a real project
- uncertainty, but to control it in such a manner as not to hide the effects of control actions, and
- opportunities for communication, working in teams and using ethical decision-making processes.

The simulation provides a mechanism for students to learn about the planning and control of a project by experiencing these processes. The simulation concerns a construction project through which students can plan and control in a real context but without the inordinate cost (such as money, time and reputation) implied by learning on a real project. Information is provided to the students about the simulation model, how it is used and the type of feedback they should expect to receive on certain aspects of their planning and control.

Therefore, the general research question posed in the study was: How effective is the dam game simulation as a quality learning and teaching tool? Specifically, it was important to make judgements about the extent to which students fostered generic graduate attributes by:

- applying scientific/mathematical and technical skills to engineering problems
- identifying, defining and providing solutions to engineering problems
- producing engineering-style reports
- structuring engineering problems into a sequence comprising identification, formulation, solution and impact
- demonstrating a range of effective engineering communication skills
- collaborating effectively in, and identifying the working benefits of, multidisciplinary and multi-cultural teams, and
- showing an appreciation of the role and responsibilities of professional engineers.

THE SIMULATION

THE PROJECT AND THE USER INTERFACE

The simulation is based on a project to construct a rock-fill dam with a clay core. The finished dam is 30 m high and 300 m wide at the top. Figure 1 shows a general arrangement of the site. It indicates some of the features to be considered by the student including the design and maintenance of temporary haul roads, the environmental impact of the work (including working close to a Site of Special Scientific Interest) and the effect of the work on neighbours. This is in addition to the normal planning and control considerations present on an isolated site. The simulation mimics an authentic construction project including planning, decision making, uncertainty, environmental effects, finance and a realistic physical model of project management and resource operation.

THE STUDENTS' TASKS

The dam game simulation is designed so that a student team adopts the role of the contractor's project manager responsible for the planning, resource selection and use, the control and reporting to the company management. Resources are required to excavate, transport and place the rock and clay and to maintain the haul roads. Figure 2 shows a typical screen for the choice of plant.

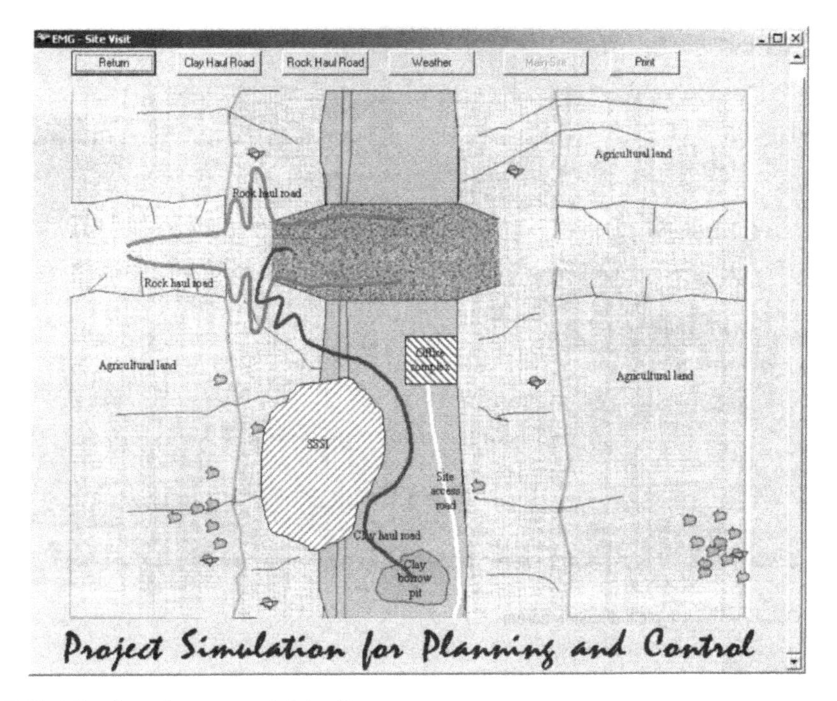

FIGURE 1 General arrangement of the site

Management resources also are required and have to be selected by the student team. For example, the numbers of engineers and foremen to supervise the rock excavation, transport and placing must be considered together with the amount of money to be spent on training them in quality, safety and environmental matters.

EXPERIENCE WITH THE SIMULATION

The simulation has been run as part of undergraduate courses in the UK, the Netherlands (Al-Jibouri and Mawdesley, 2001), New Zealand and Australia. The experience and results described here are related to the recent use of the simulation at the Civil Engineering Department at Curtin University where the simulation has been integrated into the common first year of the undergraduate course for all engineering disciplines.

THEORETICAL FRAMEWORK FOR THE USE OF THE SIMULATION

Engineering is based on understanding, analysing and shaping the real world, and while much of the process is grounded in the technical fundamentals of engineering, science and mathematics, engineering graduates also need to demonstrate that they have acquired competencies related to the more social aspects of engineering practice. Working in teams, communicating with people from diverse backgrounds and conducting themselves in an ethical and responsible way are some of these types of learning outcomes that are expected of engineering graduates. In this study, because the investigation was related to examining qualitative aspects of students' actions and interactions in a learning environment, the theoretical referent of constructivism was used to investigate the effectiveness of the simulation as a learning tool to develop learning outcomes associated with the real world of engineering.

Constructivism, while it has many forms, is a well-founded education philosophy with its roots in the seminal works of Dewey (1997), Piaget (1972) and Vygotsky (1986). Von Glasserfield (1995), a more recent proponent of constructivism in science teaching and learning, described the philosophy of education as one that 'breaks with the old concept that knowledge represents an independent world and opens the door to

FIGURE 2 A typical decision screen

examining learners and how they make sense of meaning' (p7). Modern constructivism has encompassed the idea that learning is a social and cultural act and thus learning is part of life and cannot be separated from our innate search for meaning which engages minds, emotions, feelings and attitudes. The principles that become inherent in the teaching and learning environment are those that:

- associate learners' thinking with their perceptions of the world and their language
- provide opportunities for learners to construct their understandings based on their experiences
- encourage learners to make sense through problem solving, especially through extensive dialogue with others, and
- allow learners to develop their understandings of a whole and real world context, as well as its parts.

Thus, as a theoretical referent, constructivism provides an analytical framework to investigate the effectiveness of the dam game simulation as a learning tool to develop learning outcomes associated with the real world of engineering. Real world engineering requires that students are exposed to experiential learning environments – simulations are a valuable tool to achieve this purpose.

METHODOLOGICAL CONSIDERATIONS IN THE STUDY

An interpretive model of research using qualitative and descriptive data underpinned the study because 'social reality as it is sensed, known and understood is a social production' (Denzin, 1989: 5). People hold their own local and constructed realities of their social world (Guba and Lincoln, 1994, 1998) because they are capable of shaping their own behaviour and the behaviour of others. Multiple realities exist because humans interact with one another by manipulating and negotiating symbols, language and meanings.

> *These multiple realities will invariably diverge (each inquiry raises more questions than it answers) so that prediction and control are unlikely outcomes although some level of understanding (verstehen) can be achieved. (Guba and Lincoln, 1985: 37, emphasis in original)*

The researchers did not attempt to control any variables in the study. Rather they were interested in what the students did with the simulation, how they interacted with it and what types of learning outcomes were produced because of their interactions.

The intent of the study was not to find results about the simulation that could be generalized. Rather, the strength of the study was to seek out students' perceptions of how the simulation supported their learning and how well the simulation was used as a teaching tool. The assumption taken by the teacher-researchers in the study was that reality is layered, there must be an appreciation of the culture being studied and one must try to learn and understand the symbolic meanings (the language, gestures, looks, actions and appearance) of the participants. Therefore, the explanations derived in the study were subjective. Interestingly, the researchers in the Curtin study formed a two-discipline cooperative teacher-researcher team – an engineer (the engineering content expert) and a trained teacher (the teaching and learning expert) – to allow for the checking of interpretations made about the effectiveness of the simulation as a teaching and learning tool. The teacher-researchers did not seek to compare the grades of different cohorts of students nor the relationship between the effectiveness of the simulation and student grades in any of the student groups.

As espoused by Guba and Lincoln (1994), the teacher-researchers believed that a transactional and subjective relationship needed to exist between the would-be-knowers (the teacher-researchers) and the knowers (those who participated in the study). The teacher-researchers were responsible for critically reflecting on the research process and to synthesize the quantitative and qualitative data for interpretive purposes. Choices were made about when to be involved with participants or when to keep a distance, when to create opportunities for interaction or when to stand back from events and actions and when to question possible spurious explanations. Bruner (1986) believed we are not trying to establish truth but verisimilitude in interpretive research because a study represents truly conceivable experiences. However, although there are difficulties in relying on description and story telling in the interpretive model of research, the iterative nature of the study was vital in reducing the amount of qualitative data amassed and gradually building judgements about the effectiveness of the simulation as a tool for teaching and learning, especially in relation to formative assessment.

To ensure that interpretive research is trustworthy and of good quality, Guba and Lincoln (1985, 1989) defined parallel criteria for the research standards of validity and reliability. These are the 'credibility, transferability, dependability and confirmability' of the data generated and how they are analysed (Guba and Lincoln, 1985: 43). The trustworthiness of the methodology in this study was safeguarded on several premises. The data were collected and analysed by the teacher-researchers in close liaison with the students involved in the study. The credibility (the internal validity of the study) was heightened because the research methods used in the study were triangulated. The chosen methods allowed for the development of continuous contact with students, persistent observations of them, cross-checking their perceptions through discussion and collecting and assessing students' assessment tasks. Many records were kept during the course of the study as an audit trail which included the raw data collected, data analysis results and reductions, data synthesis products and recorded notes in a journal. The journal was constantly used in the data collection periods during the study and expanded as an analytical log and reflective tool. Peer debriefing between the teacher-researchers was a valuable process to clarify interpretations made about the data. The dependability of the study was concerned with the triangulation over the 18-month period. Triangulation, the procedure where data are collected and analysed from several perspectives, is amply described in the literature (Jick, 1979; Mathison, 1988; Denzin, 1989; Denzin and Lincoln, 1998; Stake, 1998). Time, methods and data sources enhanced the validity of the study and the developmental aspect of the study added to the confirmability of the interpretations made. Using the survey method to confirm interpretations across the sample of students fostered the transferability of the findings, that is, the external validity of the study. It is conceded that further research work needs to be done to strengthen this quality criterion. Future work will focus on showing the strength of the relationship between the effectiveness of the simulation and student grades.

RESEARCH DESIGN OF THE STUDY

The interpretive model of research guided the several methods chosen for data collection, analysis and representation. These methods provided the grounding to gain the detailed, context-bound information necessary to understand first, the various ideas that students held about the simulation from their own points of view and, second, the potential to make teacher-researcher judgements about the learning and teaching processes involved in using the simulation. The information is provided in descriptive terms about how things happened and with an attempt to interpret why things happened.

The study consisted of two phases. First, a pilot study was conducted with a group of third-year civil engineers over a 12-week semester to judge how relevant the simulation was for the engineering course at Curtin. Second, a year-long study was conducted with first-year students to evaluate how readily they could engage with the simulation to learn about the plan-monitor-control process and how well the simulation allowed them to improve their communication and teamwork skills even though they did not possess a great understanding of engineering technology.

PHASE ONE: WORKING WITH THIRD-YEAR CIVIL ENGINEERS

In the third year of the civil engineering programme at Curtin University, one unit of study was concerned with engineering economics and project management. This was previously a lecture-based subject run over a 12-week semester, but in 2003 it was decided to incorporate the simulation into students' learning experiences and assessment tasks. As the students lacked experience in project management, many of the concepts (such as plan-monitor-control, risk management, cash flow and resource management) were new to them. By using the simulation, they were given the opportunity to learn how to apply theory and take on high-level responsibilities without the potential drastic results that could occur in real life. Additionally, students had to report (orally in a presentation, visually using charts and in a written project report) on plans, progress and outcomes as part of their assessment programme.

There were 53 students in the class group who organized themselves into project teams of two or three. These groups were given guidance about working as teams and the types of roles they could adopt within the project team. Participants took the role of project managers responsible for the planning and control of the project and the team approach was a beneficial strategy to facilitate discussion, peer tutoring and peer mentoring. Before starting work on constructing the project, each group was required to produce a plan of work, a financial plan and to choose a proposed method of control. These were orally presented to the 'board of directors' (the cooperative teaching team) for comments and approval. Each team was then expected to use the simulation to input the agreed plan into the computer and operate the project.

The plans produced by the students were in terms of the height of the dam against time and expenditure, and income against time. The students were reminded that they would be judged against the agreed plan as much as against the profit they made. The simulation monitored progress against this plan and attempted to make suggestions as to the need to replan. If replanning was done, the simulation monitored against both the original plan and the most up-to-date plan available.

Several qualitative research strategies for collecting data (observations, student achievement of assessment tasks and student questionnaires) were used during the research process to gather information about the effectiveness of the simulation in developing students' knowledge and professional practices. The two teacher-researchers collected the information to allow interpretations to be made about the teaching and learning processes involved in using the simulation. The strength of these interpretations was that shared understandings about students' learning converged from different perspectives (of the engineer, the teacher and the students themselves).

Observations of students working with the simulation, responding to student questions and cooperative teacher debriefings were conducted weekly over the semester period. The lectures were the main avenue to provide information to students at the appropriate stages of planning, controlling and monitoring the project. There were two main assessment points during the semester. The first was associated with the simulation and the second built on the skills gained from the exercise to enable students to develop feasibility plans for an urban development project. The cooperative teachers acted as a board of directors in both of these assessments. The assessment associated with the simulation involved three tasks – an initial set of project plans were displayed for assessment, an oral presentation of the plans showing any adjustments based on feedback from the board of directors and a written project report. Each task in the assessment was a project team effort and assessed as such.

Feedback from the students relating to the evaluation of the simulation was more formally gathered in two ways:

- a qualitative evaluation review through open-ended questions in the form of 'what have you learnt?' and
- a structured questionnaire specifically designed to evaluate the effectiveness of the simulation in teaching engineering fundamentals and management skills.

PHASE TWO: WORKING WITH COMMON FIRST-YEAR STUDENTS

At Curtin University, all first-year students, irrespective of the engineering discipline they have selected, study a foundational year to their studies (the 'common first year'). The experience in the pilot study showed that students did not acquire a number of engineering skills satisfactorily. This was either a deficiency in the syllabuses employed or the inability of traditional teaching methods to adequately address these specific areas. The decision was made to introduce the simulation to first-year students in their 'engineering professional studies' unit to provide foundation knowledge about the expectations of professional engineers in project management.

Over the next 12 months, approximately 200 students were involved in the study with the same two teacher-researchers investigating student reactions to the simulation. Based on the pilot study, several changes were made. First, the teaching approaches in using the simulation were made more explicit for the first-year students. Second, the data collection strategies were modified. Observations of students were continued as was the monitoring of students' assessment tasks, but at the end of the learning experience, students were asked to respond to four open-ended questions.

- What have you learnt about the 'plan-monitor-control' cycle in project management?
- What have you learnt about working in teams in project management?
- What have you learnt about communication in project management?
- What have you learnt about uncertainty in project management?

In addition, students were asked to rate (on a 10-point scale) the dam construction project according to how they thought it had added to their learning about project management. These changed strategies were made in an attempt to examine the specific skills students had learnt and the extent of value they placed on the simulation as a learning tool.

OUTCOMES OF THE STUDY

PHASE ONE

In the pilot phase of the study, the third-year students overall had a positive response about the effectiveness of the simulation in terms of understanding the basic principles of the design and construction process and believed that they could apply these to other situations in a practical world (see Items 1 and 2 in Figure 3).

Students indicated that they had gained a greater awareness of the planning and economic aspects of engineering. One student's response to an open-ended question exemplified other students' learning:

> *The game helped us to experience the practical knowledge of civil engineering and gain a better understanding of practical engineering. It allowed us to think of a whole project as one item instead of one part of it only.*

Students also were positive about understanding the 'plan–monitor–control' cycle in project management (see Item 4 in Figure 3). The development of this knowledge was significant and also was evident from the observations made during lecture sessions, the ways in which students tackled their assessment tasks and the positive manner in which students responded

Item	
1	I now understand the basic principles of the design/construction process.
2	I could apply basic design/construction principles in other similar representative exercises.
3	I now can use engineering judgment about the quantifiable activities of project delivery.
4	I now can use effectively the "plan-monitor-control" cycle.
5	I have learnt more about how to structure engineering problems into a sequence comprising identification, formulation, solution, and impact.

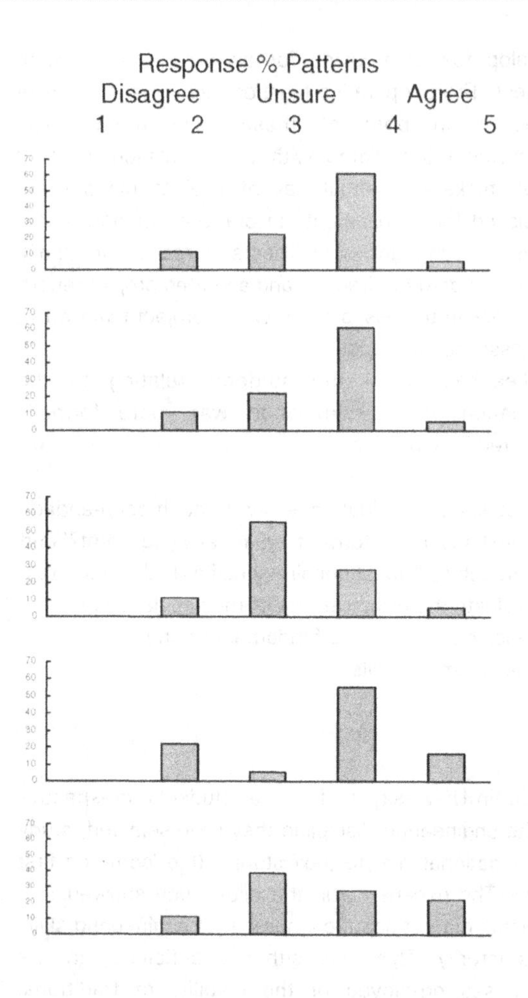

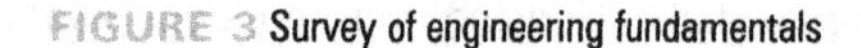
FIGURE 3 Survey of engineering fundamentals

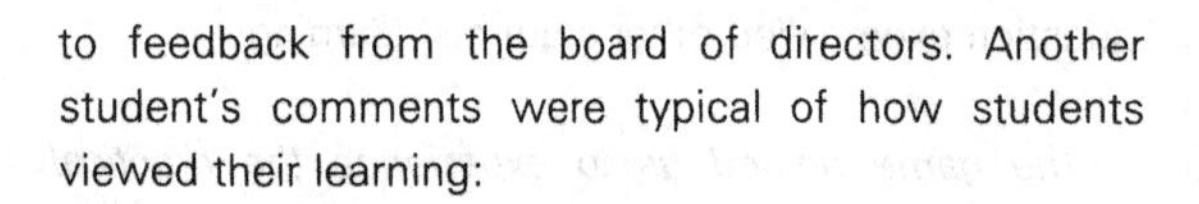

to feedback from the board of directors. Another student's comments were typical of how students viewed their learning:

It has taught me that, in engineering, original plans don't always go the way one originally thought. This was the first exercise in our course that gave me variable and unexpected outcomes (and problems).

However, there was a range of responses from students about whether the simulation improved their ability to structure engineering problems into a sequence (see Item 5 in Figure 3). Some students commented positively about their improved problem solving, as illustrated by one student:

We have learnt to identify problems and improve our problem-solving abilities. The game made us realize how one decision will affect others and that there isn't one correct answer; there are many ways to solve problems. The game has aided in improving our ability to work under pressure.

Other students were less confident about any improvement in their problem-solving abilities as noted by one student:

I don't think it helped too much. The principle behind it was good and the ideas it brought up were important but I think the game was hard to function and we were unable to correctly identify and solve problems. More background information is needed.

It appeared that engagement with the simulation did not necessarily allay several students' concerns about whether they were able to make engineering judgements concerning quantifiable activities in project management (see Item 3 in Figure 3). Some students considered that the simulation 'assisted them in a better understanding of how to use our engineering judgement' but others, as demonstrated in their assessment tasks, were daunted by the challenges embedded in quantifiable data associated with project management.

The simulation had a positive effect on students in relation to organizing a group to work as a team (see Item 3 in Figure 4). Many confirming comments by the students about the simulation were along these lines about the teamwork item.

It has provided us with the opportunity to work in a team and come to a common goal in a short period of time. It made us think through problems and make judgements as a group. We became more aware of different people types; how some are aggressive leaders and others are submissive followers. The choice of group is very important and the ability to communicate as a unity is a must.

Also, students gained an awareness of their future professional responsibilities and the impact of proper engineering procedures on the workplace, people and the environment (see Items 4 and 5 in Figure 4).

To work in a group and accept decisions of others is a real learning experience. The game has definitely made us more aware of the responsibilities of an engineer and how our engineering decisions affect a broad range of people. It has contributed in understanding the basic work environment.

While many students believed that they were more able to communicate in a professional engineering capacity (see Item 1 in Figure 4), others were unsure about their own competency with communication skills. Because the assessment of the initial oral presentation of the plans and written reports was conducted as a group assessment, it is probable that students could not gauge their individual performance as a contribution to the group. There was a range of student responses about their abilities to communicate effectively within their project team during the simulation exercise (see Item 2 in Figure 4).

The important finding from the pilot study was that students required more time to use the computer simulation without it distracting them from the task in hand. One student made this clear.

I think the dam idea was good, but I found it difficult to see how it allowed us to better understand problems. We were not using the programme properly because we didn't understand it and as a result we lost patience with it. We need to try the game beforehand to understand the value (engineering) of numbers shown, so problems are not tackled blindly. We need something like a practice game with warnings or a tutorial in the simulation which will inform the user what to expect.

Not all students had difficulty in working the simulation – as one student said, 'the game itself is good and relatively easy to understand' – but students needed to know what the computer programme required them to do in order to carry out planning, monitoring and controlling the production of the dam, i.e. the task in hand.

It became clear that the simulation did not replace the teaching of essential information required for project management; it is a learning tool that must be used in conjunction with formative and iterative teaching strategies. It was concluded that once students could master the mechanics of engaging in the simulation, their learning from the simulation could be heightened.

Importantly, the evaluation of the simulation showed that many students were frustrated because 'it was impossible to win the game, that is, make a profit!' This issue also was tied to providing more information to students about the principles of costing and making equipment and contract estimates in terms of an appropriate cost–benefit timeframe for production rather than 'winning'. It was envisaged that more information provided to students in the supporting lectures in future would overcome this problem.

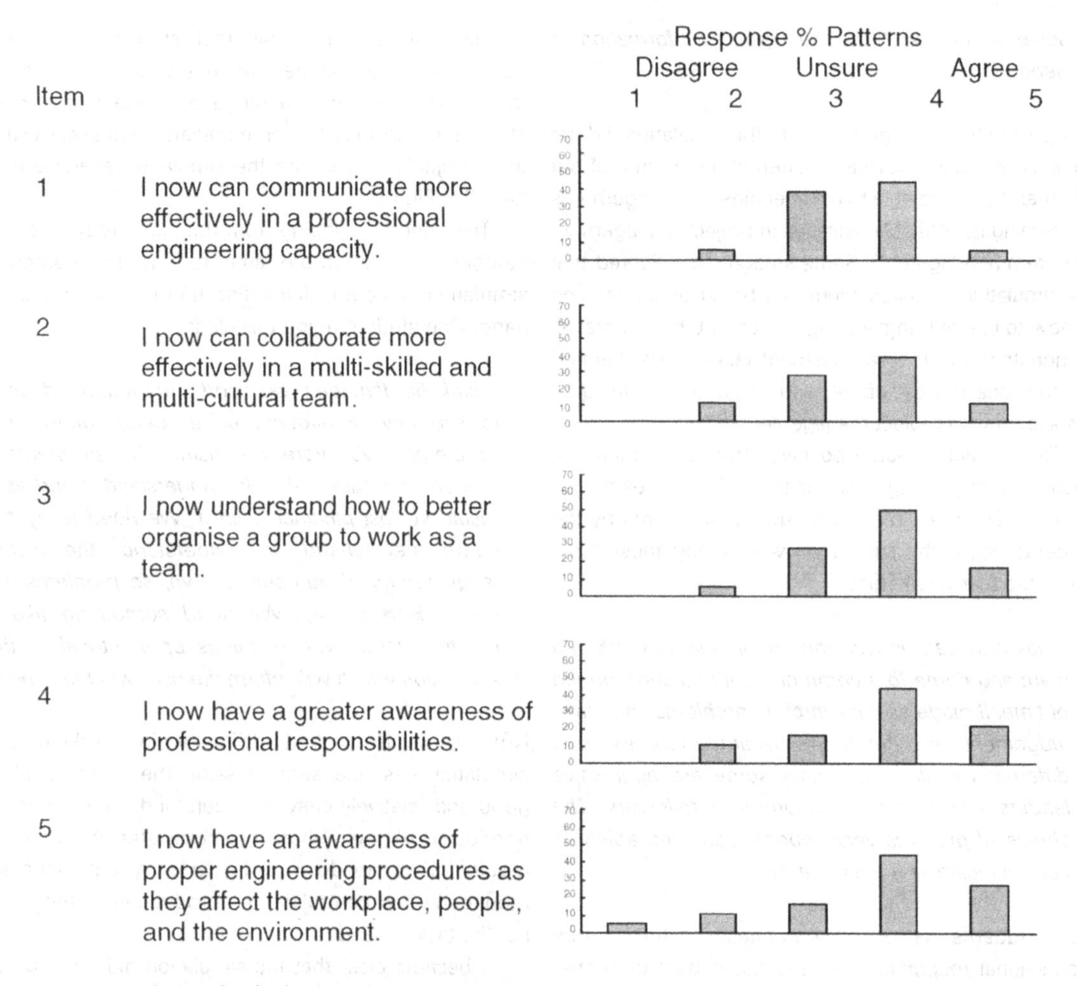

FIGURE 4 Survey of professional practice

The pilot study clearly demonstrated that the third-year students had little prior experience of working in groups as part of a learning experience. Indeed, the students commented that the team approach to learning was novel for them. While the team experience was effective for third-year students, they were still keen to know about their individual performance within their group. Some students suggested that part of the assessments could include a brief report on group efficiency in their capacity to plan, monitor and control their projects. This suggestion was worthwhile for further consideration because the strategy would assist future students to monitor and reflect on their own and others' learning.

It was revealed in the pilot study that some inadequacies existed in the level of resources available in the students' learning environment. Students required more time and more computers running the simulation to allow them greater access to both independent and group learning. Pin-up space was essential for students to adequately display their original plans (on the poster-like charts) for the review process to be conducted by either the board of directors or the students themselves. Students learn from their peers and the display of students' work was intended to be an essential part of a rich learning experience for all students. A valuable teaching and learning resource emerged from the evaluation of the simulation. Photographs were included in the formative evaluation feedback that the board of directors provided to the third-year students and signalled to students a clear rubric for the way in which their work was assessed.

PHASE TWO

In phase two of the study, the data were analyzed to produce composite narratives about students' learning experiences. The following student comments give insight into what students learnt about the plan–monitor–control process.

> *Each step in the plan–monitor–control cycle is important, especially planning. It is not easy and requires lots of hard work because the plan never equals the actual outcomes. I've never come across the plan–monitor–control cycle before in a working example (e.g. never put into practice) and I have learnt a great deal about how it works. The cycle is continuous because the site situation is continuously changing and planning must reflect the changes. Monitoring the plan helps a project team to decide what action needs to be taken to control or improve production. My team learnt that we should monitor progress regularly and control and adjust no matter what circumstances we face. This process is harder than it looks and it is a lot of work, but it is the best way to manage a project.*

Students learnt much about their teamwork skills in terms of their individual strengths and weaknesses and the contribution each made to their team effort. They learnt that working in teams to achieve a purpose was a complex process and more challenging than working in lab groups, for example, to complete a task.

> *Each team member's input is important. We are able to think on a broader scale and thus many ideas, decisions and thoughts can be expressed. There are natural born leaders who generally do all the work and that the other team members back up with information and presentation. It is important to discuss all decisions together and consider alternatives. It brings out your confidence and you learn to tolerate others and be patient with them. Working in groups can be pretty hard. We must always respect teammates' ideas and keep a high level of motivation. Cooperation is important too. Working in a team helps us to learn how a person might work under others or how to manage people working under you. It is important to make sure that the workload is distributed evenly and that the workload assigned is to the best person. Working in a team is both difficult and productive.*

In addition, students learnt that efficient and effective communication was critical to working in teams.

> *The project would be impossible to complete without keeping in contact with the other team members. Sometimes it's tough and you have to compromise. Communication allows the situation to be known by everyone in the team. It allows you to play more of an important role and also how the leadership role can be improved. Communication is the key ingredient in a successful team. Communication and understanding others, in addition to listening to others, is very important. It is the most important factor – you can't do it alone and must communicate with your peers because a team is not a team if team members do not communicate.*

Uncertainty in project management emerged as the most difficult concept for students to grasp. Students wanted to be precisely correct in their mathematical calculations and making reasonable estimates was a challenge for them. For example, a striking instance was when one project group was adamant that their expenditure for their project would be $1,245,642 and 25 cents! The following excerpt illustrates the difficulties students experienced.

> *Plan for uncertainty? It's confusing! We learnt you must always be careful about what you are going to do in the plan. Uncertainty is common as there is always a deviation from the normal plan. Things can vary a lot from the actual plan and we learnt to take into consideration other factors affecting the project. We found that we could never underestimate the error, as well as the problems that may occur. There will always*

be uncertainty. It is how we deal with it that's important. We learnt that just randomly changing multiple variables to fix problems created further uncertainty and was not cost efficient. We had to learn to problem solve for the best decision to deal with uncertainty and add a bit of common sense. Uncertainty becomes a little more predictable as you get through more and more of the project management.

A synopsis of the main points distilled from the qualitative data collected from 198 first-year students is presented in Table 1.

The overall average rating by the students during this phase of the study in relation to the value of the simulation as a learning tool was 7.5 (on a 10-point scale). While this rate was positive and students were learning about project management and developing the required graduate attributes for engineering students, there were teaching and learning issues that emerged from this phase of the study.

It remained evident that students tended to focus on learning how to use the simulation rather than developing their engineering skills. Playing the game and winning (making a profit) continued to be a prime goal for students. Separating the gaming idea from using the simulation as a learning experience was a valuable lesson learnt from the study. Furthermore, it was clear that any future teaching with the simulation first had to deal with the engineering skills to be developed so that students could create a project plan. Based on this plan, the simulation could then be used for the construction of the dam. At the time, this was not (but now is) an obvious teaching strategy. First-year students could not cope with complex multi-tasking early in their course. It was found that different tasks needed to be explicitly pulled apart, then modelled, before students could synthesize them to make use of the simulation for optimum learning. While not an issue of great concern, it was found that students also required explicit strategies to develop a team culture. More direction about how to work in teams and what different roles needed to be assigned were essential to make the best use of the simulation as part of a team project. Of greatest significance was the finding that students who use the simulation as a learning tool require constant reassurance and guidance about the uncertainty of project planning and management.

IMPLICATIONS FOR FURTHER RESEARCH

The effectiveness of the dam game simulation and others will continue to be investigated in a scientific way to demonstrate the viability of simulations and management gaming for developing some of the generic attributes required by engineering graduates. It is envisioned that the dam game simulation requires functionality to show students their patterns of engagement with the simulation and the nature of their engagement. Producing this functionality in the simulation will be beneficial to students to enable both them and their teachers to address different styles of learning. Part of any further research also will involve developing instruments that will facilitate gathering data from students in the UK, the Netherlands and Australia. The comparison of students will deepen the researchers' understanding about how students, especially in different engineering disciplines, use the simulation as a learning tool and provide direction about improving the effectiveness of simulations as an asset in engineering education.

CONCLUSION

The dam game simulation that has been used in the teaching and learning associated with planning and control of civil engineering projects has been found to be an effective teaching and learning tool for such purposes. The study has also described some of the experience gained from operating the simulation. Apart from the study, direct observation over a number of years by the authors suggests that students and professional engineers learn a great deal about many aspects of construction as a result of using the dam game simulation. The skills acquired range from plant management through to engineering cost control and effective teamwork. This observation is endorsed by the enthusiastic support and participation of industry groups.

However, the study has shown that further research work still is required to inform engineering educators about how to use the simulation for even better learning to be gained from its use. In order to formally demonstrate the usefulness and effectiveness of the

TABLE 1 Student opinions on the game

QUESTION	TYPICAL RESPONSES
What have you learnt about the 'plan–monitor–control' cycle in project management? It is a very important cycle and all steps must be followed; continue to alter your plans, monitor these plans and control this, reassessing the plans when things start going wrong. A good project needs to have a detailed plan to be monitored continuously and for actions to be made to these plans and progress. Without careful planning control will be difficult. Monitoring also is essential. To monitor and control sections which should be weighted far more heavily as planning is subject to rather large uncertainty and hence variance.	Each step is important especially planning.
What have you learnt about working in teams in project management? Anything is possible when everyone cooperates, commits to the tasks, and works together – more ideas flow and alternatives can be considered so that the best results are achieved. There are born leaders in teams and the others back up the team with ideas and carrying out tasks so that the workload is distributed evenly and the workload is assigned to the best person. Teamwork brings out the confidence of each member and everyone learns to listen, be patient and show tolerance and respect for team members. Teamwork is hard work.	Each team member is equally important.
What have you learnt about communication in project management? Communication allows the situation to be known by everyone in the team. Communication is the key ingredient in a successful team. Communication is essential for success especially if each member is dealing with a separate aspect of the project. You learn to be in a group/team, deal with conflict and differences in opinions and to make decisions.	The project would be impossible without keeping in contact with the other members.
What have you learnt about uncertainty in project management? Things can vary a lot from the original plan and you must learn to take into consideration other factors affecting the project. I have learned to be prepared for the worst to come out of uncertainty; it is stressful. There always will be uncertainty; it is how we deal with it – that's what is important. We have to plan and take into account the unexpected problem.	Uncertainty is common as there is always a deviation from the normal plan.

dam game, a systematic evaluation of learning needs will be initiated across larger student data sets. The effectiveness of the simulation to improve student grades is a matter for further research.

AUTHOR CONTACT DETAILS

Susan J. Gribble (contact author): Division of Engineering, Science and Computing, Curtin University of Technology, GPO Box 1987, Perth, Western Australia 6845, Australia.
Tel: +61 8 9266 3890, fax: +61 8 9266 2606,
e-mail: J.Gribble@exchange.curtin.edu.au
David Scott: Department of Civil Engineering, Curtin University of Technology, GPO Box 1987, Perth, Western Australia 6845, Australia.
Tel: +61 8 9266 7573, fax: +61 8 9266 2681,
e-mail: D.Scott@curtin.edu.au
Mick Mawdesley: Nottingham Centre for Infrastructure, School of Civil Engineering, University of Nottingham, University Park, Nottingham NG7 2RD, UK.
Tel: +44 (0) 115 951 3897, fax: +44 (0) 115 951 3898,
e-mail: Michael.Mawdesley@nottingham.ac.uk
Saad Al-Jibouri: Construction Management and Engineering Group, Department of Civil Engineering, University of Twente, PO Box 217, 7500 AE Enschede, The Netherlands.
Tel: +31-(0) 53-4894887, fax: +31-(0) 53-4892511,
e-mail: S.H.Al-Jibouri@sms.utwente.nl

REFERENCES

Al-Jibouri, S. and Mawdesley, M., 2001, 'Design and experience with a computer game for teaching construction project planning and control', in *Engineering Construction and Architectural Management*, 8(3/4), 418–427.

Au, T. and Parti, E., 1969, 'Building construction game – general description', in *Journal of the Construction Division, ASCE* 95(CO1), 1–9.

Bruner, J., 1986, *Actual Minds, Possible Worlds*, Cambridge, MA, Harvard University Press.

Denzin, N.K., 1989, *The Research Act: A Theoretical Introduction to Sociological Methods*, 3rd edn, Englewood Cliffs, NJ, Prentice Hall.

Denzin, N.K. and Lincoln, Y.S., 1998, 'Introduction: entering the field of qualitative research' in N.K. Denzin and Y.S. Lincoln (eds), *Strategies of Qualitative Inquiry*, Thousand Oaks, CA, Sage Publications.

Dewey, J., 1997, *Experience and Education*, Riverside, NJ, Simon & Schuster (original work published in 1938).

Gilgeous, V. and D'Çruz, M., 1996, 'A study of business and management games', in *Management Development Review*, 9(1), 32–40.

Guba, E.G. and Lincoln, Y.S., 1985, *Naturalistic Inquiry*, Newbury Park, CA, Sage Publications.

Guba, E.G. and Lincoln, Y.S., 1989, *Fourth Generation Evaluation*, Newbury Park, CA, Sage Publications.

Guba, E.G. and Lincoln, Y.S., 1994, 'Competing paradigms in qualitative research' in N.K. Denzin and Y.S. Lincoln (eds), *Handbook of Qualitative Research*, Thousand Oaks, CA, Sage Publications.

Guba, E.G. and Lincoln, Y.S., 1998, 'Competing paradigms in qualitative research' in N.K. Denzin and Y.S. Lincoln (eds), *The Landscape of Qualitative Research: Theories and Issues*, Thousand Oaks, CA, Sage Publications.

Jick, T.D., 1979, 'Mixing qualitative and quantitative methods: triangulation in action', in *Administrative Science Quarterly*, 24(4), 602–610.

Mathison, S., 1998, 'Why triangulate?', in *Educational Researcher*, 17(2), 13–17.

Piaget, J., 1972, *Psychology and Epistemology: Towards a Theory of Knowledge*, Harmondsworth, Penguin.

Sawhney, A., Mund, A. and Koczenasz, J., 2001, 'Internet-based interactive construction management learning system', in *Journal of Construction Education*, 6(3), 124–138.

Scott, D. and Cullingford, G., 1973, 'Scheduling games for construction industry training', in *Journal of the Construction Division*, 99(CO6), 81–92.

Stake, R.E., 1998, 'Case studies' in N.K. Denzin and Y.S. Lincoln (eds), *Strategies of Qualitative Inquiry*, Thousand Oaks, CA, Sage Publications.

Tommelein, I., Riley, D.R. and Howell, G.A., 1999, 'Parade game: impact of work flow variability on trade performance', in *Journal of Construction Engineering and Management, ASCE*, 125(5), 304–310.

Von Glasserfield, E.A., 1995, 'Constructivist approach to teaching', in L.P. Steff and J. Gale (eds), *Constructivism in Education*, Hillsdale, NJ, Lawrence Erlbaum.

Vygotsky, L.S., 1986, *Thought and Language*, A. Kozulin (translator and editor), Cambridge, MA, The MIT Press.

Wankat, P. and Oreovicz, F., 2004, 'Simulators to stimulate students', in *Prism*, 13(5), 45

Wiesner, T.F. and Lan, W., 2004, 'Comparison of student learning in physical and simulated unit operations experiments', in *Journal of Engineering Education*, 93(3), 195–210.

ARTICLE

Using Video in the Construction Technology Classroom

Encouraging Active Learning

Mike Hoxley and Richard Rowsell

Abstract

During the past 15 years, the use of video in the classroom at all levels of education has increased while, at the same time, most research into educational technology has concentrated on personal computers and the Internet. Consequently, there is a lack of research into how video is used in teaching at a time when it is one of the most used technologies. What research has been carried out (mainly in the medical education domain) has generally found video to be effective in promoting student learning and that students are receptive to its use. However, it is necessary to ensure that students engage in active (rather than passive) viewing. This paper reports the authors' experience of using the materials produced by the Video Project at the University of the West of England (UWE) in teaching level 1 domestic-scale construction technology at Anglia Ruskin University. The research is concerned with how the videos may best be used in the lecture theatre. Data, collected by questionnaire from more than 200 students, largely support the authors' approach of using a short but carefully focused quiz as an 'orienting activity' to encourage 'active learning'. Feedback of the quiz results can then be used as the means by which further detail and reinforcement of key points is provided.

■ *Keywords* – Construction; learning; teaching; technology; video

INTRODUCTION

> *Video is uniquely suited to take students on impossible field trips – inside the human body, or off to Jupiter.*

This somewhat extravagant claim is made by a US public service broadcaster in its educational video promotional literature (Thirteen Edonline, undated). In the UK, video has long been used in the classroom in schools (Moss *et al*, 1991), further education (BBC/SFEU, 1994) and in higher education (Barford and Weston, 1997). Houston (2000) reports that as the use of video in the classroom has grown, research into its use has waned. This is because research activity in instructional technology has shifted to media such as personal computers and the Internet. Using the case study of the authors' institution, this paper investigates the *effective* use of video in facilitating student learning of domestic construction technology at level 1 of HE.

The research reported in this paper is not concerned with Web-based video for use by students in a computer lab or at home, but with the use of video in the classroom. Web-based technologies can of course be used to supplement more traditional styles of teaching (for example, see Shelbourn *et al*, 2004) but given that a majority of UK universities own the videos produced by the University of the West of England (UWE), this research has explored the most effective way of using these learning materials in the lecture theatre.

WHY USE VIDEO?

It has been suggested that, in the teaching of the 'TV generation', fundamentally different strategies need to be adopted than hitherto. Gioia and Brass (1985–86) observe that students have grown up 'in an intensive environment of television, movies and video games' and have developed learning styles where comprehension occurs largely through visual images. They warn of a mismatch of traditional teaching methods, lecture and textbook readings, and the visual learning styles of contemporary students. Certainly the authors are surprised by the admissions that their students make concerning their lack of recreational reading – although there are noticeable gender differences, with female students generally reading much more than males.

On the other hand, books have been characterized by Kozma (1991) as a learning medium for their 'stability'. The stability of the written word offers several advantages to the learner. It enables the reader to control the rate at which information is received and those with highly developed reading skills can skim read at their own pace. The authors are not advocating that video should *replace* reading as the only learning medium, but that it can be used to enhance the learning process as just one weapon in the armoury. The other 'weapons' are described later in this paper.

Using videos in higher education has been shown to both improve students' examination marks and reduce tutorial support time (Rae, 1993). Marx and Frost (1998) provide a comprehensive review of the use of video in management education and suggest that video can convey meaning that is difficult to match with traditional lecture and reading assignments. They report that management educators have been impressed by the ability of video to engage students and managers. However, Gioia and Brass (1985–86) warn against catering to bad habits of reinforcing learning modes that support passive, superficial consumption of video offerings rather than the more desirable offerings that define higher education. Rogow (1997) cautions against 'using the television as a babysitter'.

Video is particularly useful in two types of situation – where some *technique* needs to be demonstrated and where students require a *visual appreciation* to understand (Meisel, 1998). Obviously both of these situations apply in the study of construction technology. Meisel states the following rules for the use of teaching and training videos:

- Never show a video of someone else saying what you can say.
- Use videos for things you cannot adequately describe (e.g. emotions, broad application of theory to practice, etc.).
- The absolute need to prepare and plan. This is to get beyond the audience reaction characterized by: 'It was a great show but I don't know if I learned anything'.

Demonstrating techniques is important in the medical professions and several studies have been undertaken to assess the effectiveness of the use of video compared with other teaching methods in medical education. In a study of teaching clinical skills in assessing and managing drug-seeking patients, three methods were compared (Taverner *et al*, 2000). Small group tutorials, video-based tutorials and computer-aided instruction (CAI) were used to teach the same skills to different groups of students within the same cohort of senior medical students over two years. The CAI development costs were higher, but there was no significant difference in the results of assessment for the three groups. However, the students preferred the video-based tutorials to the other two methods. Similarly, Dequeker and Jaspaert (1998) found that video-supported small-group learning of problem solving and clinical reasoning 'can promote enjoyable learning for students and teachers'. In a study of orthodontic auxiliary training, video teaching was found to slightly out-perform a slide-based lecture, in the training of the placing of orthodontic brackets (Chen *et al*, 1998). Parkin and Dogra's (2000) experience of using video in undergraduate teaching of child psychiatry was that 93% of 249 students found videos 'useful' or 'very useful' in learning about assessment and disorders in child psychiatry.

Given that the evidence presented above suggests that video-based teaching is at worst *no less* effective than more traditional methods and that students seem to prefer it as a method, then this seems a good enough recommendation to adopt video-based learning where, of course, it is appropriate.

HOW TO USE VIDEO IN THE LECTURE THEATRE

As we have seen above, when reading, the learner is able to control the pace at which they learn, whereas the pace of video 'is not sensitive to the cognitive constraints of the learner; it progresses whether or not comprehension is achieved' (Kozma, 1991). Kozma suggests that learning from video occurs through a 'window of cognitive engagement' which refers to the visual attention that learners focus on the video's content. Many advocates of video use in the classroom have encouraged an 'active viewing approach' rather than a 'passive viewing approach' (Wetzel *et al*, 1994). Kreiner (1997) suggests that guided note-taking of video material may improve learning compared with passive observation. The use of 'orienting activities' is advocated by Hooper and Hannafin (1991) and these can include stating lesson objectives before showing the video. Rogow (1997) recommends 'use of the board or overhead projector to write out a few questions relating to the video. Go over the questions before running the tape, so students will know what to look for.' Houston (2000) carried out a questionnaire survey of more than 500 community college faculty members in the US. She concluded that the use of *active learning* strategies in the classroom is one way to reduce students' tendencies to view videos passively and increase student participation in the learning process.

Marx and Frost (1998) see the greatest challenge of using video as 'harnessing the motivating impact of video without falling prey to its failings – shallow comprehension, trivialization, and lowered mental effort'. They advocate meshing video and printed learning materials for optimal educational outcomes. Such an approach has been adopted by the developers of the Video Project at UWE.

UWE'S VIDEO PROJECT

This project was established in the early 1990s and has produced more than 20 films on building construction and building conservation in a UK context (Marshall, 2001). UWE claims that over half of the built environment courses in the UK use these videos, and at UWE they are used to teach architects, planners, construction managers, surveyors, housing managers, estate agents and environmental health officers. The videos may be used as 'stand alone' elements or as a part of a wider lecture and tutorial package. Each video is approximately 25 minutes long, combines site/factory footage, professional narration and high-quality graphics. There is a tutorial workbook to accompany most videos and students complete the workbooks as part of their directed study. The video and workbooks are supported by a textbook (Marshall and Worthing, 2000). This fully integrated learning package is highly acclaimed throughout the UK and the considerable contribution of the UWE project developers to the study of building technology is gratefully acknowledged by the writers of this paper.

ANGLIA RUSKIN UNIVERSITY'S EXPERIENCE

Following use of the complete learning package by one of the authors for two years at another institution, Anglia Ruskin University adopted it for the first time in 2000/01. We therefore have five years' experience of its use. In the Department of the Built Environment, the building technology and services module is taught in the first year of all courses (architecture, civil engineering, construction management and surveying). There are typically 120 part-time students and 80 full-time students studying the module and the different modes are taught separately. A formal lecture to each group (during which a video is usually shown) is followed by tutorials in groups of about 20 where students work in groups of three or four to complete, discuss and mark the workbooks, and also engage in other practical exercises. At the time that data for this paper were collected, the authors were module leader (Rowsell) and deputy module leader (Hoxley) and were assisted by three other teaching staff in the delivery of the module. We firmly believe that this first module in building technology is of crucial importance to students. All of the built environment professions to which students aspire, require a sound grasp of the technology that underpins them. As well as knowledge and understanding of construction functional requirements and processes, there is a whole new vocabulary to learn.

An additional set of videos is available in the reference section of the learning resource centre, so that a student missing a lecture is able to view it later and any student is able to view a video for a second time should they wish to do so. The advantages of using videos in teaching construction technology are well understood – they reduce the need for (and therefore the risk of) site visits, students can view processes not easily communicated in a formal lecture and the subject takes on a real live dimension. However, we still take students on one site visit during the module and believe that this experience is invaluable, particularly for full-time students who may never have visited a construction site before. The module has 48 hours of classroom contact time and a recommended 152 hours of self-study. It is assessed by an assignment and an end of module examination. More recently, the completion of a minimum number of the tutorial workbooks has been made compulsory.

From the outset of using the videos, we saw the need to maintain students' attention during the playing of the video and to encourage 'active' rather than 'passive' learning. The 'orienting activity' we have adopted is that recommended by Rogow (1997) – to get students to complete a short quiz during the playing of the video. Our first attempt at this was a disaster! The quiz was far too long and students spent more time looking down at the quiz than watching the screen. We then tried showing the quiz on an overhead projector at the same time as the video was playing. However, students said that they would prefer the quiz on a handout that they could spend a few minutes studying before the video played. We have experimented with the number of questions but believe 10–12 to be about right.

TO LECTURE OR NOT TO LECTURE?

The amount of formal lecturing required in addition to the showing of the video depends on the level of detail required for a particular topic. We have tried lecturing before and after the video (see results of data collection below) but a solution we have found to be quite effective with some topics is to make the feedback of the quiz answers the focus of learning. If this method is to be adopted then the quiz requires very careful design to ensure that the questions asked are the key points of the topic. Feedback on the answers to the quiz is then used as the vehicle for providing the necessary further detail and to reinforce these key points.

THE STUDENTS' PERCEPTION

Towards the end of the second and third sessions of using the videos, we sought feedback from students in order to help inform any necessary changes. A two-page questionnaire was given to all students attending a lecture in the penultimate week of the module. A total of 209 completed questionnaires were returned and a blank questionnaire may be seen at Appendix A.

The breakdown of the student sample by course studied is shown in Figure 1. There were 131 part-time and 78 full-time students in the sample.

206 students (98.6%) supported the concept of showing videos. Even if the two missing responses are taken as being negative, this result is a clear vote of approval for the use of video to teach level 1 construction technology. One civil engineering student commented: 'I really enjoy the videos. Books and lecturing are useful but to actually see work in progress is very beneficial. Definitely keep with the videos, they're useful.' An HND property and surveying student said: 'I found this subject particularly interesting, the use of videos is a fantastic idea as I do not get the opportunity to visit a working site very often.'

The reasons given for liking videos are illustrated in Figure 2 (respondents were able to indicate more than one reason). 'Demonstrating site processes' and 'preferring to watch a video than read a book' were the highest responses at 108 and 103, respectively. The latter response is rather alarming but perhaps not surprising given that most students are part of Gioia and Brass's (1985–86) 'TV generation' – and indeed mostly male!

189 students approved of the use of a quiz with only 16 saying that they did not. The reasons given are illustrated in Figure 3 and again students were advised that they could give more than one reason. 'Reinforcing the main points' scores more highly than 'easier to remember' and 'aiding concentration'. Perhaps this result is yet another indication that what really focuses a student's attention is assessment. It also stresses the importance of the tutor ensuring that the quiz questions do indeed focus on the key issues. 15 of the 16 students who did not think the quiz was a good idea stated that they found it a distraction and some said that they would prefer to make their own notes. We suspect that these students are all 'high performers' but, since the data collection was carried out anonymously, further research would be required to confirm this view.

Only five students preferred the quiz to be on an overhead slide with 201 preferring a hand-out (there were three missing responses). The number of quiz

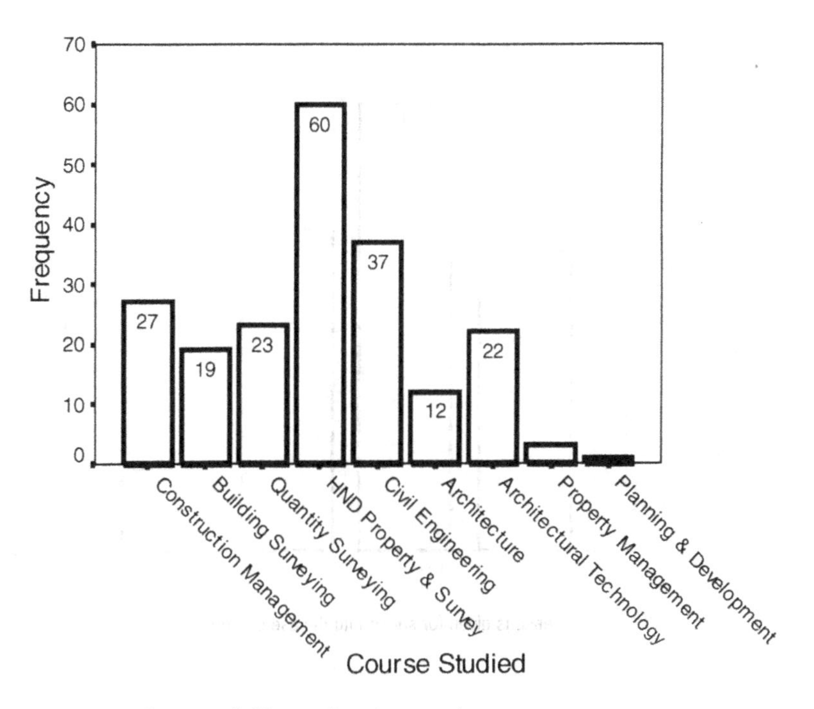

FIGURE 1 Course studied by questionnaire respondents

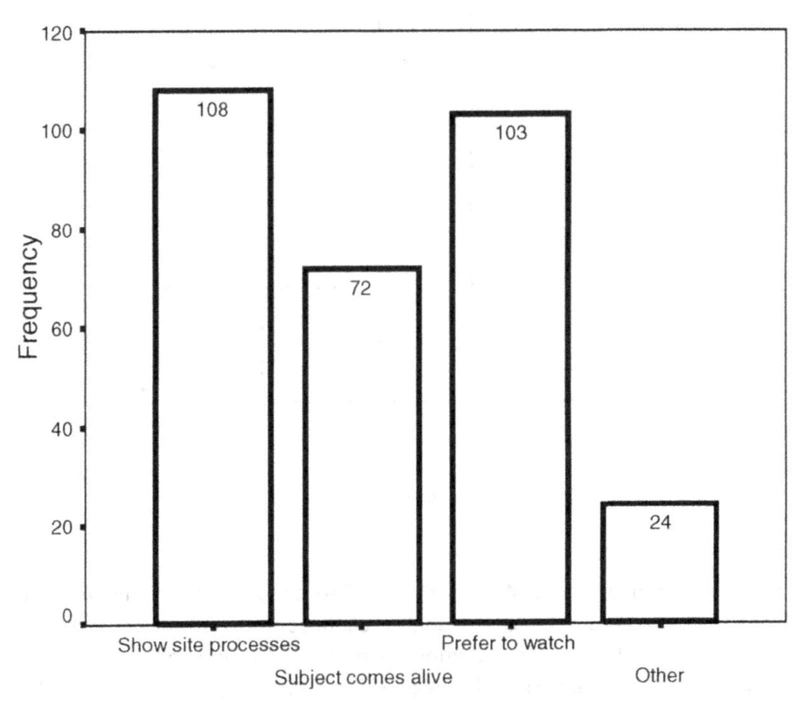

FIGURE 2 Reasons given for supporting the use of videos

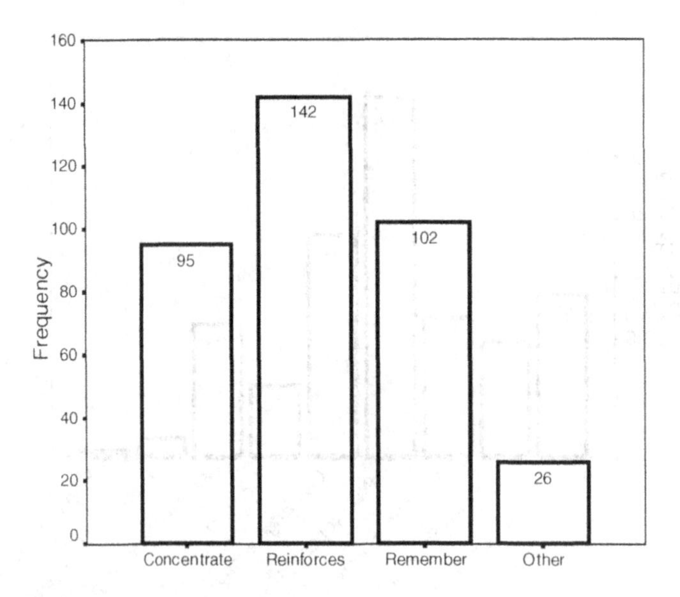

FIGURE 3 **Reasons given for supporting the use of a quiz**

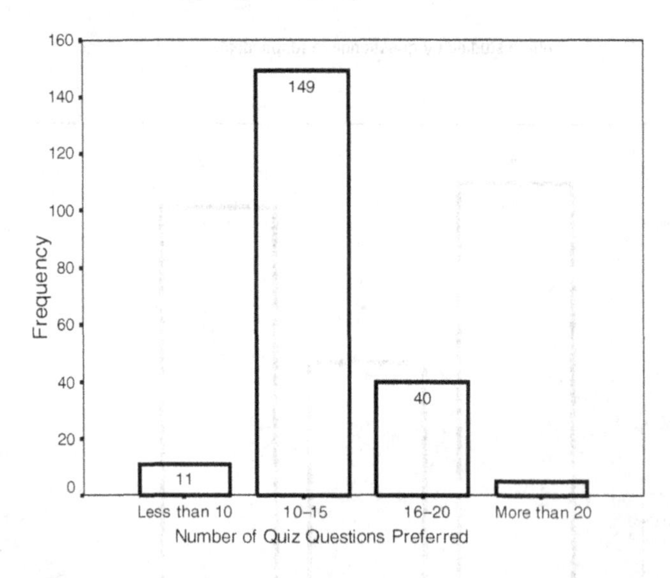

FIGURE 4 **Preferred ranges of quiz questions**

questions preferred is indicated in Figure 4 and the authors' view (given above) seems to have been endorsed by these cohorts. Only 13 students admitted to borrowing a video from the library and nine of these said that it was because they had missed a lecture.

The area of most disparity with the authors' preconceptions was about whether a formal lecture was required in addition to the video and quiz feedback. 152 students said that a traditional lecture was required (49 thought that it was not). There is also strong support for having the lecture after the video (171 students) rather than before (34 students).

Having reflected on these results, the authors believe that where the lecture introduces new concepts, provides an overview of a topic, or discusses functional requirements, then it is probably best to lecture before the video. Where, however, the main purpose of the lecture is to deliver technical detail, then this is certainly best delivered after the viewing of the video. This approach, informed by student feedback, is the one that we have since adopted.

CONCLUSIONS

The use of video to teach technology to built environment undergraduate students is now a common feature of UK courses. This may be partially in response to a desire to adapt teaching methods to cater for students brought up in an age of TV, but it is also, no doubt, because an excellent learning package has been made available by the Video Project of UWE. Many institutions have realized that there is no point in 'reinventing the wheel' and that the UWE videos, tutorial workbooks and accompanying textbook provide a fully integrated system for delivery of a domestic-scale construction technology course at a reasonable price. The authors' experience of using video in this context is that there is a need to aid students' concentration while showing the video and to encourage 'active' rather than 'passive' viewing. The use of a short quiz provides this aid and also usefully allows tutors to move seamlessly into lecture mode during feedback of the quiz solutions. However, data collected from students as part of this study suggest that there is still a need for the traditional lecture and that this is usually best delivered *after* the video has been shown. A small minority of students (7%) said that they found the use of the quiz a distraction. These may be high-performing students but further research would be required to confirm this view.

AUTHOR CONTACT DETAILS

Mike Hoxley and Richard Rowsell: Department of Built Environment, Anglia Ruskin University, Bishop Hall Lane, Chelmsford, Essex, CM1 1SQ, UK. Tel: +44 (0) 1245 493131 ext 3082, fax: +44 (0) 1245 252646, e-mail: m.hoxley@anglia.ac.uk

REFERENCES

Barford, J. and Weston, C., 1997, 'The use of video as a teaching resource in a new university', in *British Journal of Educational Technology*, 28(1), 40–50.

BBC/SFEU, 1994, *Report on the Impact of Broadcasting in Further Education Colleges in Scotland*, BBC Scotland and the Scottish Further Education Unit.

Chen, M., Horrocks, E. and Evans, R., 1998, 'Video versus lecture: effective alternatives for orthodontic auxiliary training', in *British Journal of Orthodontics*, 25(3), 191–195.

Dequeker, J. and Jaspaert, R., 1998, 'Teaching problem-solving and clinical reasoning: 20 years' experience with video-supported small-group learning', in *Medical Education*, 32(4), 384–389.

Gioia, D. and Brass, D., 1985–86, 'Teaching the TV generation: the case for observational learning', in *Organisational Behavior Teaching Review*, 10(2), 11–18.

Hooper, S. and Hannafin, M., 1991, 'Psychological perspectives on emerging instructional technologies: a critical analysis', in *Educational Psychologist*, 26(1), 69–95.

Houston, C., 2000, 'Video usage and active learning strategies among community college faculty members', in *Community College Journal of Research and Practice*, 24(5), 341–357.

Kozma, R., 1991, 'Learning with media', in *Review of Educational Research*, 61(2), 179–211.

Kreiner, D.S., 1997, 'Guided notes and interactive methods of teaching with videotapes,' in *Teaching of Psychology*, 24(3), 183–185.

Marshall, D., 2001, 'Video learning resources for building technology', http://www.be.coventry.ac.uk/BPBNetwork/casestudy/westeng_lre1i.htm (accessed 1 March 2005).

Marshall, D. and Worthing, D., 2000, *The Construction of Houses*, London, Estates Gazette.

Marx, R. and Frost, P., 1998, 'Towards optimal use of video in management education: examining the evidence', in *Journal of Management Development*, 17(4), 243–250.

Meisel, S., 1998, 'Videotypes: considerations for effective use of video in teaching and training', in *Journal of Management Development*, 17(4), 251–258.

Moss, R., Gunter, B. and Jones, C., 1991, *Television in Schools*, London, John Libbey (for the Independent Television Commission).

Parkin, R., and Dogra, N., 2000, 'Making videos for medical undergraduate teaching in child psychiatry: the development, use and perceived effectiveness of structured videotapes of clinical material for use by medical students in child psychiatry,' in *Medical Teacher*, 22(6), 568–571.

Rae, A., 1993, 'Self-paced learning with video for undergraduates: a multi-media Keller Plan', in *British Journal of Educational Technology*, 24(1), 43–51.

Rogow, F., 1997, 'Don't turn off the lights: tips for classroom use of ITV', http://www.scetv.org/education/k-12/resources/classroom_tv.cfm (accessed 1 March 2005).

Shelbourn, M., Hoxley, M. and Aouad, G., 2004, 'Learning building pathology using computers – evaluation of a prototype application', in *Structural Survey*, 22(1), 30–38.

Taverner, D., Dodding, C. and White, J., 2000, 'Comparison of methods for teaching clinical skills in assessing and managing drug-seeking patients', in *Medical Education*, 34(4), 285–291.

Thirteen Edonline, undated, 'Video strategies', http://www.thirteen.org/edonline/ntti/resources/video1.html (accessed 1 March 2005).

Wetzel, C.D., Radtke, P.H. and Stern, H.W., 1994, *Instructional Effectiveness of Video Media*, New Jersey, Hillsdate.

APPENDIX A

Student questionnaire on the use of videos in building technology and services

1. Which course are you on?

   ....

2. Do you think that the videos are a good idea? (please tick)

   ❑ Yes ❑ No

3. If you said No, please say why below; if you said Yes then please tick one or more boxes below to say why you like videos.

   ❑ They demonstrate site processes without the need to visit site
   ❑ They make the subject 'come alive'
   ❑ I prefer to watch the video than read the book
   ❑ Other (please state below)

   ....
   ....
   ....

4. Do you think that a quiz for you to complete as you watch the video is a good idea?

   ❑ Yes ❑ No

5. Why? (again you may tick more than one box)

   ❑ It helps me to concentrate
   ❑ It reinforces the main points
   ❑ I find it easier to remember what I have watched
   ❑ Other (please state below)

   ....
   ....
   ....

6. Do you prefer the quiz on a handout rather than just on the overhead projector?

   ❑ Yes ❑ No

7. What do you think is about the right number of questions for the quiz? (please tick one box)

   ❑ Less than 10
   ❑ 10–15
   ❑ 16–20
   ❑ More than 20

8. Have you borrowed any of the videos from the library?

   ❑ Yes ❑ No

9. Why? (you may tick more than one box)

   ❑ Because I missed a lecture.
   ❑ To reinforce what I saw in the lecture.
   ❑ Other (please state below)

   ....
   ....
   ....

In the remaining questions, please think about the use of the videos in comparison with a 'traditional lecture'.

10. Is the video best before or after the lecture on the subject?

    ❑ Before
    ❑ After
    ❑ No lecture required

11. If the feedback to the quiz is comprehensive, do you think that you still need a 'traditional lecture' on the subject?

    ❑ Yes ❑ No

Many thanks for your help. We would appreciate any other comments you have on the videos – please write them on the other side of this page.

Richard and Mike

ARTICLE

Developing Web-based Tools for Teaching, Training, Learning and Development

The Role of Academic Institutions

Mohan M. Kumaraswamy, Christopher J. Miller, M. Motiar Rahman, David G. Pickernell, S. Thomas Ng and Israel P.Y. Wong

Abstract

Increasing demands and decreasing resources in the built environment sector point to critical needs for more effective teaching and learning tools for academia, as well as more efficient individual training and organizational learning mechanisms for industry. This paper 'overviews' two Web-based tools that target the upgrading of teaching-learning-training opportunities of university students and small and medium contractors (SMCs), respectively. Although these developments are based in Hong Kong, the overviews are preceded by a review of various challenges facing students and SMCs and the need for information and communication technology (ICT)-aided enterprise education in general. In terms of undergraduate teaching-learning, the reduced opportunities for actual site visits are being compensated for through Web-based 'virtual site visits', while supplementary knowledge on specific topics, such as 'construction work study', have also been captured in CIVCAL – a computer-aided teaching – learning package for civil engineering, building and construction undergraduates. The needs of SMCs in Hong Kong are being prioritized and addressed by the academia-led SMILE-SMC project, which it is envisaged will be eventually accessed by 'large' contractors, as well as consultants and clients. This would then enable them to communicate seamlessly on project-specific information management platforms to improve performance on specific projects and, indeed, across more sustainable construction supply chains.

■ *Keywords* – Internet; teaching; training; knowledge; Hong Kong; information management; academic institutions

INTRODUCTION

Many secondary school leavers consider the prospects of an engineering and construction (EC)-related programme for their tertiary study. Given recent infrastructure booms, EC programmes have been one of the favourite choices among the higher education curricula in China and Hong Kong. Although many tertiary educational institutions around the world have been offering EC programmes for years, approaches to teaching EC subjects remain largely unchanged, the most commonly used method of delivery to date being face-to-face lectures.

Some successful examples of evolution in EC education have emerged in Hong Kong, and it is valuable to draw from these and feed into selected positive moves for mutual reinforcement and synergistic development. For example, these may be compared with both a) recent high-profile shortfalls in Web-learning initiatives and 'e-university failures' (Masslen, 2004) and b) well-structured analyses of e-learning and pedagogical challenges in construction management, together with proposed solutions for 'a blended learning approach' between academia and industry (Wall and Ahmed, 2004).

The need for innovative approaches can also be seen in the construction industry itself. Construction industry improvements have been suggested through improvements of some eight critical success factors (CSFs) (Abraham, 2003). Four of these CSFs are organization-specific:

- organizational structure (the form in which an organization is internally structured to carry out its business)
- technical applications (the use of technical applications for advancement of the company)
- employee enhancements (the lifelong learning process for employees tied to personal, professional and cultural growth) and
- process benchmarking (the identification and comparison of processes and procedures and their continual improvement).

These issues are based within the organization and can be addressed through internal organizational or cultural changes – mainly through training and improving personnel capability of adapting and using new technologies and strategies – in order to perform competitively in the industry (Rahman *et al*, 2003). These are also related to, at least to some extent, information technology (IT) and information management (IM). Therefore, effective IM and IT may be considered a means to enhance project success and also to improve the competitiveness of construction organizations, thereby accelerating construction industry development.

Small and medium contractors (SMCs) are ideally positioned to benefit from collectively generated IM and IT enhancements. Having mostly operated in a 'hand-to-mouth' mode, they cannot invest or divert their scarce individual resources to searching and researching such issues (Rahman *et al*, 2004a). They work as main contractors on small projects and subcontractors/suppliers on large projects (Rahman *et al*, 2004b). Their work may constitute up to 90% of the total value of the project (Matthews *et al*, 1997). They form the largest number of business organizations in the construction sector, employ the largest number of employees and make up the largest share of the industry (Kaplan, 1996; Rahman *et al*, 2004a). Their broader significance in Hong Kong, for example, may be appreciated in that the construction industry here was recently noted to employ more than 9% of the workforce, contributes about 6% to the gross domestic product (GDP) and accounts for 40% of gross domestic fixed capital formation (CIRC, 2001), although these 'percentage contribution numbers' have now declined with decreasing construction demand.

The background data indicating the large SMC sector (as mentioned above) suggest that improvements in the construction industry generally cannot be achieved without focusing specifically on SMCs. Furthermore, any industry improvement initiative needs collective learning and knowledge transmission/sharing – to the extent that an organization requires it, and this requirement can be more for SMCs. Such collective learning and knowledge exchange is fostered by cultural, institutional and geographical proximities, often in combination (Rahman *et al*, 2004b). The experience of 'regional clusters' of high-technology small and medium enterprises (SMEs) in Europe also indicates that '[business] networks and [organizational] dependencies may be within, between and outside firms and although they may not be traded [/exchanged] (or even tradable [/exchangeable]) they may have significant effects on the competitive performance of organizations' (Keeble and Wilkinson, 1999).

There thus appears an ideal opportunity to bring together higher education in EC, SMCs and relevant IT and IM to formulate synergized solutions. In this paper, some challenges to EC education are highlighted, followed by an overview of how these relate to the challenges faced by the construction industry. The paper then discusses two examples of innovative Web-based initiatives in Hong Kong, with one addressing the teaching and learning needs of undergraduates, and the other focusing on the learning and training needs of SMCs and their personnel.

ENTERPRISE EDUCATION AND PERCEIVED BENEFITS TO THE CONSTRUCTION INDUSTRY

Research indicates that skills deficiencies exist in smaller enterprises in areas such as strategy, planning, marketing and sales and identifies a lack of formal vocational training culture within SMEs (Miller *et al*, 2002).

Gibb (1998) suggests that the dominant mode of learning in SMEs is experiential, namely, learning by doing. This includes learning from customers and suppliers, also from problem solving, opportunity taking and learning by error. It is also suggested that informal unplanned education via social and business networks is important to small firms.

In addition, it is commonly agreed that knowledge will become increasingly important in sustaining a nation's competitive advantage (Packham and Miller, 2000). We live in a digital age, the speed of technological advancement is transforming our society and therefore it should not be surprising that this technology has the potential to revolutionize training and learning (Ravet and Layte, 2001: 2). An online Web-based system can:

> *Track an individual's progress on their career plans or against targets for the whole organization ... it is the most practical and cost effective way to update skills and share information between large numbers of people. (Hammond, 2000)*

For both organizations and individuals, some of the main benefits can be summarized as follows:

- cost savings (Fry, 2001; Clarke and Hermens, 2001)
- increased access to training (Fry, 2001; Clarke and Hermens, 2001)
- flexible and continual learning (Clarke and Hermens, 2001)
- knowledge on demand (Thorne and Mackay, 2001).

The Internet can be used to simply transmit Web-based training materials to the users' computers, to be used offline by downloading the course materials. Alternatively, the Web can be used as an online instructional medium itself. However, the introduction of online methods entails a sharp learning curve for the teacher as well. When distanced from their students, the teacher can often feel isolated (Benfield, 2000). In a classroom, the teacher faces an initial struggle to establish an environment of free communication with every new class. Online, it is necessary to establish a comfortable 'computer-mediated communication' facility.

Raelin (2001) also found that online learning technology could inhibit action (or research-based) learning because of the absence of non-verbal and socio-emotional transmitted information. However, Hiltz *et al*, (2000) presented evidence that while learning in isolation online may be less motivating than learning in a traditional classroom, working collaboratively online may actually lead to higher motivation than from within a traditional classroom setting. Canning (2002) also found that even when online facilities were established at work for online delivery of Web-based materials, learners preferred to actually undertake their learning at home. It is not surprising that such preferences (to work online) are heightened in construction industry practitioners, who are usually subject to long working hours and frequent relocation.

Hughes *et al* (2002) suggest that for online collaboration to be most effective, participants must see the value of expending the (considerable) effort required, be comfortable with and trust the medium, their instructor (or facilitator) and their fellow collaborators, and feel as though they are immersed in a rich, engaging and rewarding social experience. Jones and Martinez (2001) found that, compared with the general student population, students choosing Web-based distance-learning courses tend to have learning orientations characterized by more self-directedness and discovery learning, which is in line with the dominant form of learning in SMEs. Some individuals may be attracted to distance learning because it offers them an opportunity to learn autonomously and effectively without having to interact much with others. This may be especially true in the case of busy industry personnel, who are drawn to distance learning because they do not have time to take traditional courses. Again, this can be even more applicable to SMC personnel, who often have longer and more irregular working hours.

While some technical issues remain that influence the speed of transmission of materials – especially large graphics or video files, for example – the Web is capable of effectively delivering text, graphics, animation, video and audio. The above discussion confirms that the Internet can also be seen as a potential medium to effectively address some of the deficiencies in construction industry skills, knowledge and training facilities, outlined below.

PROBLEMS FACED BY SMALL CONSTRUCTION FIRMS AND RELATED CHALLENGES FOR THE INDUSTRY

The construction sector generally has a significant small firm bias, as discussed in the Introduction e.g. in terms of SMCs possibly carrying out up to 90% of the value of work on a construction project. Despite the importance of small construction enterprises in general, and SMCs in particular, the construction sector generally (including in the UK and Hong Kong) has placed very little emphasis on the development of SMCs, seeing it merely as a part of the fragmentation of the industry, due to the structure of demand (Morton and Jagger, 1995). Large contracting firms have instead concentrated on core activities and subcontracted peripheral activities to SMCs. Small firms, however, have failed to improve technical competency and efficiency and have experienced problems that 'affect the delivery of a competent service to clients' (Cox and Townsend, 1998). These are particularly apparent for SMCs that do not benefit from the economies of larger-scale firms.

Indeed, Hall (1991, 1995) submits that small firms within the construction industry should be coerced into adopting technologies and practices that promote the development of a skilled workforce and effective project management. Briscoe *et al* (2001) have identified that the take-up of contemporary initiatives and generic skill development continue to constrain small firm development in the industry.

The issue of training initiatives is also crucially relevant to SMCs, as recognized by Stewart *et al* (2003), where it was suggested that craftsmen-entrepreneurs tended to concentrate on continuing their trade skills in business rather than dealing with management and administrative duties. Stewart *et al* (2003) also stated that craftsmen had less experience in marketing and sales. These are core skills that would be most beneficial to SMCs and they are skills that need harnessing and developing.

In their empirical investigation into management skills in the construction SMC sector, Briscoe *et al* (2001) found that 'the problems of skill inadequacy are most acute among the small and medium-sized enterprises (SMEs) that dominate the industry's structure'. As SMCs make up such a significant proportion of the construction industry it is, therefore, essential that these skills gaps are effectively addressed.

The Latham report (1994) showed how low productivity in construction, low levels of skills and relatively low investment in labour training, resulted in poor performance. The Egan report (1998) followed on from the Latham report and advocated modern supply-chain management principles as well as integrating project management techniques in the UK construction industry. The report also stated that 'better generic management skills would be required to bring about significant change in the industry' (Briscoe *et al*, 2001).

Similar findings and recommendations are set out in a high-level report on the Hong Kong construction industry (CIRC, 2001). For example, special sections were devoted to 'nurturing a professional workforce' (incorporating 16 specific recommendations) and developing an 'efficient, innovative and productive industry', with 19 specific recommendations, that for example, included one for 'developing a common platform for electronic communications within the local construction industry' and another on the wider adoption of information technology (IT). Implementing such recommendations requires an IT-literate SMC community, given that SMCs constitute a very large chunk of the industry.

ADDRESSING INDUSTRY CHALLENGES FROM ACADEMIA

GENERAL

The challenges of EC education reflect to a large extent the need to address the challenges to the industry described above. As shown in Figure 1, the construction industry is characterized by technological advancements, dynamic and heterogeneous projects, uncertainties, complexities and multidisciplinary teams. EC programmes need to equip students with lifelong skills to critically analyse the problems and derive creative solutions together with other project team members.

Many state-of-the-art techniques targeting higher productivity in EC have evolved over many years. Educational bodies should expose students to not just the basic principles but also the front-end knowledge and cutting-edge technologies when designing the curricula. Besides, as EC problems are usually quite dynamic, the critical-thinking abilities and problem-solving skills of graduates are essential for success in their future careers. Soft and qualitative knowledge and skills are now needed to complement the hard-factor knowledge

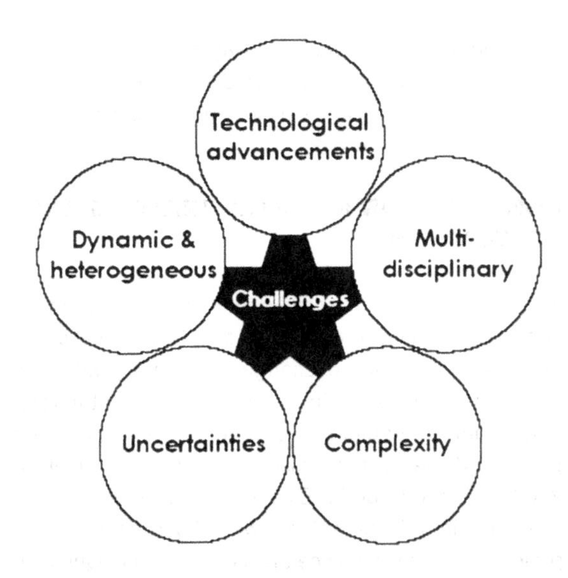

FIGURE 1 Challenges facing the construction industry

and technical and quantitative skills that engineers are identified with. For instance, EC curricula now include project management, construction law and contract administration, and students who do not have any working experience may find it difficult to comprehend the intensive knowledge relating to these areas.

Communication skills are also becoming more important, as we need to work with, coordinate and sometimes even compete with other professions who have integrated high-level management skills into their training and development programmes. Students should be exposed to the latest in information and communication technologies (ICT) which not only advance EC theory and practice but also help to accelerate communication at project and organizational levels.

EXAMPLES FROM HONG KONG

Many innovative teaching and learning methods have been introduced to the EC curricula by various institutions to prepare students for the challenges of the construction industry. Taking the Department of Civil Engineering at The University of Hong Kong (hereafter referred to as the department) as an example, local and overseas site visits as well as summer training are organized for undergraduates to empower them to learn from 'real' practice. Additionally, in-class and inter-university competitions are set up to stimulate students' understanding and creative thinking abilities.

The rapid advances in ICT and the change of learning patterns/habits of students, suggest the imperative for mobilizing the full potential of Web-based delivery. Led by the Department of Civil Engineering of The University of Hong Kong, four universities in Hong Kong that offer EC programmes (i.e. The University of Hong Kong, Hong Kong University of Science and Technology, City University of Hong Kong and The Hong Kong Polytechnic University) jointly developed a Web-based teaching – learning package known as CIVCAL.

The CIVCAL package was initially conceived to 'bring sites to the students', since it was becoming increasingly difficult to take large classes of students to construction sites. Apart from overcoming such difficulties through 'virtual site visits', CIVCAL 'captures' and 'provides on demand' special scenes and valuable details from completed projects. It also mobilizes multimedia tools in a Web-based teaching – learning package to illustrate theory and demonstrate useful applications in chosen topics, such as for improving productivity in construction processes through construction work study.

The integrative framework, formats and common search and access facilities were developed by The University of Hong Kong-led sub-team. However, in recognition of the different priorities and desired presentation formats, the opening page of CIVCAL is designed for separate access to each university domain through four distinct university-specific gateways. Each university domain reflects the particular needs of their own type of EC students e.g. the City University domain focuses more on practical building and construction scenarios, whereas others cover various civil engineering disciplines and topics.

The University of Hong Kong (HKU) opening page indicates a matrix structure as shown in Figure 2. This enables access either:

- through one of 13 projects as listed across the top row of the first page (as in Figure 1) or one of 12 topics as listed across the top row of the second page (e.g. 'water supply in Hong Kong') or
- through one of the six disciplines (ranging from 'construction' to 'water') in the first column to the left of either page (as also seen in Figure 2).

As shown in Figure 2, the projects include the ongoing Three Gorges Dam megaproject in Mainland China, and range from the famous Hong Kong Airport to repair projects on infamous landslides, and also from major highway projects to traditional high-rise and prefabricated building construction. Topics, on the other hand, range from geology, transportation and waste water treatment in a Hong Kong context, to the soil-nailing technique for slope protection, and construction work study.

Plug-in IT tools such as QuickTime movies, Shockwave (Flash), Acrobat Reader and Whip Autocad drawing viewer are mobilized to enhance various presentations. For example, a QuickTime panorama of the BioSciences Building site provides a useful overview, while Flash animations help visualize construction techniques and sequences in the Three Gorges Dam, Hung Hom Bypass and the Steep Cut Construction projects. Programmes, charts and drawings are viewed more conveniently through Acrobat Reader and Whip Autocad Drawing viewer, while video clips effectively track and demonstrate construction activities in, for example, the construction work study module.

Figure 3 indicates an example of the thumbnail sketches at the second level of the multi-level structure in each module. Each of these thumbnails would lead to a full image with description/narration/illustration/animation. The extensive multimedia database includes photographic narratives, construction details, work programmes, video clips and animations, as well as navigation and teaching tools developed specifically for CIVCAL. This package can therefore boost both classroom teaching and post-lecture learning activities.

CONSTRUCTION INDUSTRY TRAINING AND DEVELOPMENT

IDENTIFIED NEEDS

In order to meet market demands, production or construction management must also take into account and optimize the exchange of data with the environment of the firm, i.e. clients, contractors, consultants, suppliers, subcontractors and other partners (Caillaud and Passemad, 2001; Ugwu *et al*, 2003). IT can provide SMEs in general with appropriate solutions in this regard as well. Each construction project is unique and since SMCs play a very significant role in the construction industry, SMCs must optimize the use both of their resources and their exchanges of internal and external data through IM (Rahman *et al*, 2004b). Caillaud and Passemad (2001) state, in manufacturing scenarios, that IM is important to:

- clarify the strategy by gathering micro- and macro-economic intelligence
- anticipate new 'product' developments by a technological intelligence and a study of the political and legal tendencies (constraints, subventions) and

FIGURE 2 Opening page in the 'HKU domain' with top row tabs leading to 'projects'

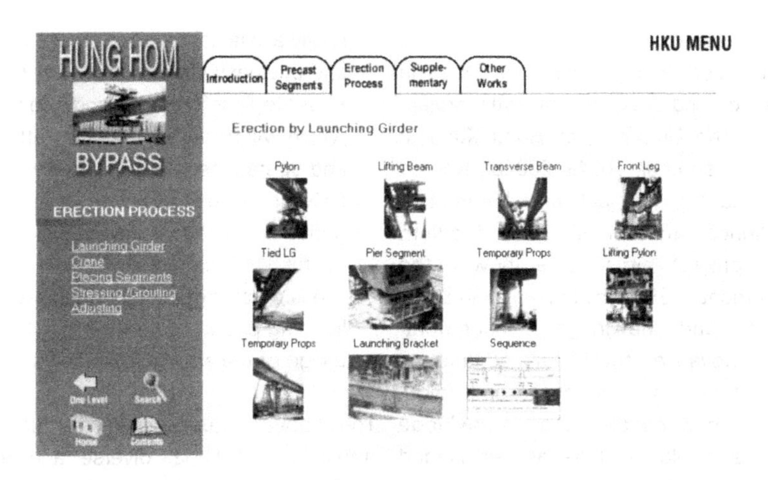

FIGURE 3 Thumbnail images that can be 'opened' to illustrate a project 'process'

- acquire a flexible, reactive and communicating organization ... continuous information about the 'production' processes and the transverse management of the information circuits, through the optimization of its exchanges of information with the environment, the evaluation of its performances and through the capitalization of its knowledge.

In construction scenarios, just as in many other sectors such as manufacturing, SMCs must integrate these various dimensions by using tools in accordance with their needs, such as the management of various processes (through workflow management), the sharing of data/experience (through a common database or platform) and the capitalization of experience (through knowledge management (KM)). In the case of distributed networks (i.e. the supply chain in a construction scenario), a single-point information source (e.g. data warehouse), expert systems (e.g. various generic process-related modules that can be used by a specific organization) and other decision support systems can be more effective (Stevens, 1999). A framework of an information system based on the Internet and other available IT tools answers these needs and allows continuity between the external and internal information flows through the coexistence of heterogeneous and distributed systems (Caillaud and Passemad, 2001). From the perspectives of SMCs, the areas where effective IM (aiming at faster and smoother decision making) can be applied with the help of the Internet and other IT tools include (Rahman *et al*, 2004b):

- Data and information required by construction site staff from head office: foremen/supervisors, gang leaders, individual trades people (e.g. plasterers) – in smaller projects with no site office.
- Data and information required by construction site staff from site office: foremen/supervisors, gang leaders, individual trades people (e.g. scaffolders).
- Data and information required by site office from various sources: foremen/supervisors, gang leaders working at site, individual trades people working at site (e.g. plasterers), head office, client/owner/developer, consultant, subcontractor(s) and supplier(s), other supplier(s) working on the same project/site (if any).
- Data and information required by head office from various sources: 1) in smaller projects with no site office – a) foremen/supervisors working at construction sites and b) individual trades people working at sites (e.g. plasterers and scaffolders); 2) different site offices; 3) clients/owners/developers, consultants, main contractors (if any), subcontractor(s) and supplier(s); 4) other subcontractor(s) working on common projects/sites (if any), other supplier(s) working on common projects/sites (if any); and 5) other external organizations, e.g. Labour Department, etc.

OVERVIEW OF SMILE-SMC

Addressing the identified specific needs of SMCs in Hong Kong, a research and development (R&D) project was launched from The University of Hong Kong in November 2003 (Rahman *et al*, 2004a, 2004b). Named SMILE-SMC (Strategic Management with Information Leveraged Excellence for Small and Medium Contractors), the project aims to empower the continuous improvement of SMCs by providing an SMC-friendly 'information and knowledge management' framework and innovative tools for continuous improvement in boosting productivity, quality and image, through strategic information and knowledge management. For example, savings are envisaged through reduced wastage (of resources) and less rework (in rectifying substandard or defective works). The objectives of the SMILE-SMC project were stated as:

- to identify, consolidate and develop good practices, critical success factors and appropriate benchmarks for SMCs
- to develop a comprehensive framework and innovative tools for enhancing SMC competitiveness through structured modelling, knowledge capture, information management, collaborative team working and benchmarking mechanisms
- to develop a SMC-friendly Web-based 'strategic information and knowledge manager' to empower improved productivity, quality, safety and other critical performance aspects
- to implement (on a 'pilot run' basis in at least three SMCs), test and refine the above framework and tools.

The above objectives are targeted using 'easy to use' and available IT tools and through a 'one-stop' information source. The planned deliverables of the SMILE-SMC project are:

- A Web-based information library for boosting SMCs' business and operations. This will include information on new technologies, business opportunities and innovative approaches for SMCs.
- A collaborative information and knowledge management framework that captures (and makes easily available) relevant SMC strategic information and business process knowledge.
- A 'strategic information and knowledge manager', which will be an SMC-friendly affordable business and operations support-cum-advisory system for SMCs to include templates and initial issues of periodical (e.g. quarterly) newsletters/e-bulletins on useful SMC matters.
- A basic training workshop package and a self-learning package to enable quick and effective usage of the above deliverables.

The above requires the collection, distillation and translation of huge, diverse and hitherto scattered packets of information – multi-project, multi-organizational and organization-specific data for and from SMCs – with sources varying from those needed for their day-to-day business activities to strategic decision making. Therefore, close collaboration and active participation from a motivated set of pioneering SMCs or partner contractors (PCs) was felt to be essential. 16 such PCs were enlisted in stages, along with collaborating organizations, such as the Construction Industry Training Authority, Hong Kong Construction Association, General Building Contractors Association, and Hong Kong Construction Subcontractors Association. The PCs supply required information, assist in collecting further information from their trade/sub-partners/contractors, attend monthly full-team meetings and provide feedback and suggestions in developing the SMILE-SMC system, and will participate in its validation and dissemination.

Advice is also obtained from overseas advisers in five institutions: two based in the UK, two in Australia and one in Singapore. The adviser in Singapore is at the National University of Singapore, while those in Australia are at the WeB (Working for eBusiness) Centre at Edith Cowan University and the ACCI (Australian Centre for Construction Innovation) at the University of New South Wales. The UK advisers are at the CICE (Centre for Innovative Construction Engineering) and the WEI (Welsh Enterprise Institute) at the University of Glamorgan.

To develop the above-mentioned deliverables, a pilot 'needs analysis' exercise was conducted through semi-structured interviews with the first 14 PCs, thereby eliciting their problems, needs, ideas and suggestions.

The exercise also identified priorities on various kind of information, barriers to obtaining that information, issues of their concerns that they would like to discuss on a common platform, automation/ enhancement of various work/business processes they think helpful for improving SMC competitiveness, and the like.

On the whole, the exercise confirmed the need for the type of services planned for SMILE-SMC. All the information and constructive suggestions collected from the needs analysis exercise were discussed and distilled with the PCs and other collaborators. It was decided to offer the SMILE-SMC services in five broad zones as shown in Figures 4 and 5 – 'wanted zone', 'available zone', 'information library', 'discussion forum' and 'performance improvement'. Other important features of the SMILE-SMC website include the 'member zone', 'news and events', a section for 'search' and a section for 'downloads'. These are briefly described below.

Wanted zone

This zone is provided to facilitate a part of SMC business-related activities. SMILE-SMC members can solicit any relevant services through this zone, e.g. in looking for a business partner, contractor, subcontractor or supplier; asking for any equipment, materials, people (such as engineer, quantity surveyor, carpenter or plasterer) or other information (e.g. on contractual matters). Although only the SMILE-SMC members will be entitled to post messages here, non-members will be allowed to visit this zone and respond directly if appropriate.

Available zone

Given that most SMCs frequently work on small-scale projects, their resource mobilization is also relatively low. Yet, leftover/unused materials are a common phenomenon in construction. SMCs cannot always dispose of such leftovers economically, leading to wastage. This zone is planned to provide SMILE-SMC members with a facility to advertise such availabilities and offer any other services, for example to be a subcontractor or supplier, wishing to buy unused materials or hire in/out any equipment and any other available resources. Although only SMILE-SMC members will be authorized to post their messages offering any services, non-members may also visit this zone and respond directly if appropriate.

Information library

The main contents of the information library are not planned to be restricted to any potential user. Anyone can search this zone for any available information.

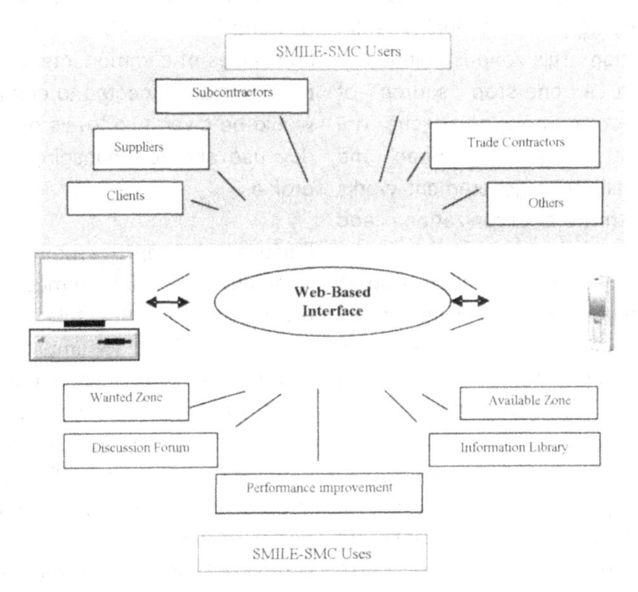

FIGURE 4 Potential users and uses of SMILE-SMC system

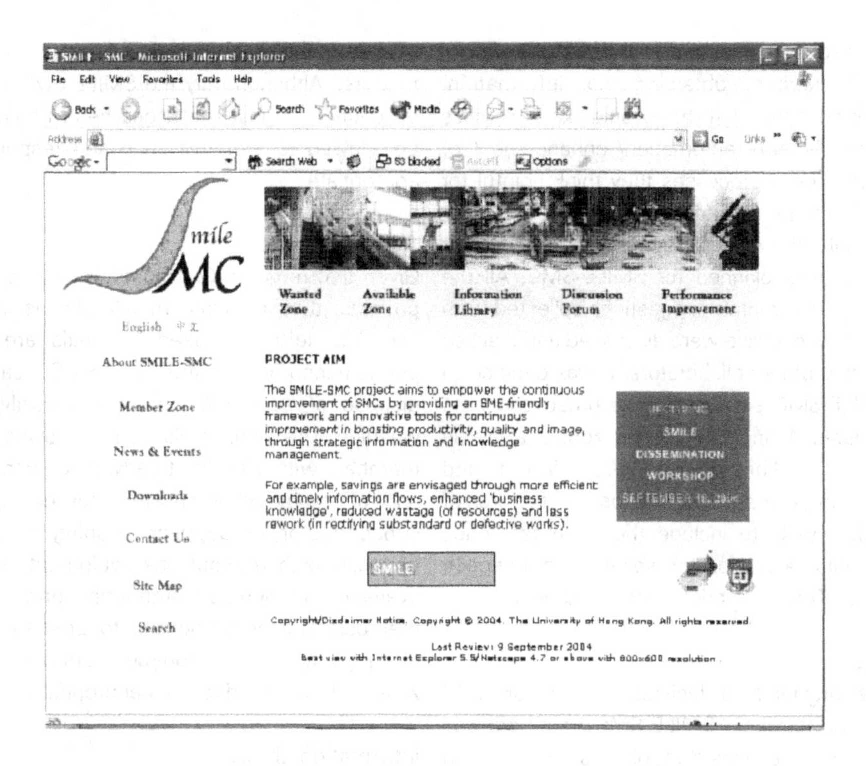

FIGURE 5 Screenshot of SMILE-SMC homepage

Information is being categorized under five broad groupings of 'statutory or regulatory information', 'business practices', 'strategic information', 'useful forms' and 'other information'. This zone is planned to eventually develop into a one-stop source of information for SMCs. For example, this zone will provide a comprehensive listing of all 'open' and 'upcoming' tenders from all major government works departments, quasi-government organizations and major private clients.

Important and frequently used information is being consolidated and presented in easy-to-understand and user-friendly formats with flow charts and pointers wherever possible, for example, on 'how to obtain a noise permit'. Examples are being provided in appropriate cases. In addition, various terms and examples are being hyperlinked both within the SMILE-SMC website and to any other related or parent website. Such formats, examples and hyperlinks are being provided and identified with the close cooperation of the PCs – the actual users.

Discussion forum

This zone is being developed in order to provide a platform for the SMCs to discuss any special concerns and issues of common interests. Only the SMILE-SMC members are expected to enjoy this service. Members would be given two levels of access here – for day-to-day use and for managing/updating their own user profiles.

Performance improvement

Contents of this site are mostly for members. Services provided through this site include generic formats/templates for improving day-to-day business activities, strategic management of critical information and knowledge, and benchmarking. Services under this zone are being developed in three modules – electronic information exchange (EIE), strategic information and knowledge management (SIKM), and benchmarking. Some of the services (e.g. templates) can be for company specific uses, for example, management of sensitive/confidential information and knowledge that is a part of SIKM. On the other hand, some other services would require members to disclose certain selected company specific information to chosen partners, for example, in benchmarking.

As a part of the EIE, a plug-in is being developed that can be used by a group of people, for example, in the project environment and by the contractor, subcontractor, supplier, consultant and client. An attempt is being made to provide a free platform authentication. Inter-organizational benchmarking is planned to be provided under a special 'benchmarking' module of this zone, whereas intra-organizational benchmarking is being provided for within the SIKM module. However, some of the services (e.g. templates) will be common to both intra- and inter-organizational benchmarking.

Other services

At the outset, the 'member zone' presents the application procedure to become a member of SMILE-SMC in an easy to understand flowchart. After joining, SMILE-SMC members can update their member profiles and user accounts. However, anyone can search for any member from this site. The 'news and events' section is planned to cover industry news and events related to SMCs. This will include, for example, news on any important training that is suitable for SMCs and changes in the rules and regulations. A 'search facility' is being provided to search both within the SMILE-SMC website and access the Internet. Moreover, anyone will be able to download any relevant document/information from a selected cross-section of documents/information.

Many parts of the SMILE-SMC website are being developed in a bilingual mode – in English and Chinese, with critical parts already translated into Chinese in order to meet specific 'local' requirements. Meanwhile, it has been planned to accelerate easy access to the SMILE-SMC services through 'training workshops', apart from periodical newsletters and e-bulletins. In addition, a self-learning package is being developed to enable potential SMILE-SMC users a quick and easier path to early usage.

From the above, it may be discerned that the various SMILE-SMC zones can serve multiple purposes, for example, by providing support tools for expanding business opportunities and improving business processes, as well as rapidly expanding the hitherto limited learning opportunities, tools and materials for both SMC organizations and their personnel. Still, some SMILE-SMC facilities may eventually be found to be more suitable/successful for enhancing learning than others. For example, the 'information library' could be developed into a popular online one-stop information source, whereas the 'discussion forum' empowers interactions, information interchange and learning opportunities. The 'performance improvement' zone provides specific modules, with tools and templates for training SMC personnel in areas of practical importance such as purchase control and materials management. It is noteworthy that these modules are being developed by academia in a classical R&D mode i.e. a structured basis of initial needs analysis, workflow analyses and feedback from industry followed by refinement, at each stage of development. This project thus provides a good example of bridging the oft-lamented gap between industry and academia, while directly and indirectly promoting and empowering e-learning and practical training in industry.

CONCLUSIONS

In response to future demands and requirements of the construction industry, educational institutions should take a more proactive role in ensuring that those who graduate from EC programmes have a strong competitive edge. In the past, it has been said that students in Hong Kong have been allegedly 'not "active" enough' in the learning process. In view of the changing and complex nature of construction, there is a need to equip students with lifelong learning and problem-solving skills and many tertiary educational institutions have already introduced some problem-based elements and other innovative approaches in teaching and learning. However, the success of EC students also depends very much on the support of the industry. A much closer industry – academia link would certainly help higher education in setting priorities in their curricula and moulding students to face the challenges of the construction industry.

The CIVCAL package provides a good example of innovative Web-based tools for enhanced teaching – learning at university level, through a vastly expanded knowledge-building environment, by reaching beyond the traditional lecture room e.g. through virtual visits to interesting construction scenarios. CIVCAL has fulfilled its expected role in supplementing traditional teaching and learning tools. Furthermore, CIVCAL-type packages may soon be seen as almost necessary to attract and sustain the interest of an Internet-linked generation of students.

On the other hand, the SMILE-SMC package exemplifies potential industry – academia linkage benefits through the research-based development of ICT-based information and knowledge management systems, for significantly expanding the learning and training opportunities of industry practitioners and organizations themselves. The need for such initiatives is justified by the background provided in this paper e.g. on the challenges facing the construction industry in general and the potential benefits from ICT-enhanced enterprise education and training in particular.

The next stage requires further evaluation and refinement of these Web-based tools, based on feedback on usage levels, usability and usefulness, as well as changing preferences and priorities. Informal surveys and regular feedback from both students and teachers have already conveyed that CIVCAL has served (and is serving) its intended purpose but a more structured survey is planned, for example, to discern more about how it contributes to a) a more effective learning environment, b) increasing student understanding and c) how it can be improved upon further.

The SMILE-SMC services are still under development. A series of workshops has already been held – between June 2004 and July 2005 (a development workshop, two dissemination workshops and a training workshop) – to present and obtain intermediate feedback from a wide cross-section of the industry for guiding further developments and refinements in addressing the needs and priorities of SMCs. A wide range of needs and priorities have been and are being identified from the feedback and from continuous collaboration with industry representatives on the SMILE-SMC team as well. These inputs are being consolidated and fed into system improvements before opening access to the full industry. Larger contractors, consultants and clients have already expressed interest in accessing and using the SMILE-SMC platform.

When completed by March 2006, the SMILE-SMC initiative is expected to make a mark in its contributions to the continuous learning and training of industry practitioners and organizations. This will be incentivized by attracting both organizations and their officers to the various SMILE-SMC facilities that can help them to improve their competitiveness and deliver effective and efficient projects while targeting value for money. Practical learning, training and knowledge-building should follow. Therefore, taking CIVCAL and SMILE-SMC together, the wider teaching, learning and training opportunities for EC students and industry are also expected to collectively contribute to, if not accelerate, construction industry development.

AUTHOR CONTACT DETAILS

Mohan M. Kumaraswamy, M. Motiar Rahman, S. Thomas Ng and Israel P.Y. Wong: Centre for Infrastructure and Construction Industry Development, The University of Hong Kong, Hong Kong SAR, China. Fax: +852 2559 5337, e-mail:mohan@hku.hk

Christopher J. Miller and David G. Pickernell: Welsh Enterprise Institute, University of Glamorgan, Pontypridd, Wales, CF37 1DL, UK. Tel: +44 1443 482380, fax: +44 1443 482380, e-mail: cjmiller@glam.ac.uk, dgpicker@glam.ac.uk

REFERENCES

Abraham, G.L., 2003, 'Critical success factors for the construction industry', in K.R. Molenaar and P.S. Chinowsky (eds), *Proceedings of ASCE Construction Research Congress, Hawaii, USA, 19–21 March 2003*, Reston, VA, ASCE Construction Institute, Construction Research Council, CD Rom.

Benfield, G., 2000, 'Teaching on the web – exploring the meanings of silence', *Ultibase Articles Online, 2000 Edition, RMIT*, http://ultibase.rmit.edu.au/Articles/online/benfield1.pdf (accessed July 2005).

Briscoe, G., Dainty, A.R.J. and Millett, S., 2001, 'Construction supply chain partnerships: skill, knowledge and attitudinal requirements', in *European Journal of Purchasing & Supply Chain Management*, 7(4), 243–255.

Caillaud, E. and Passemad, C., 2001, 'CIM and virtual enterprise: a case study in a SME', in *International Journal of Computer Integrated Manufacturing*, 14(2), 168–174.

Canning, R., 2002, 'Distance or Dis-stancing Education? A case study in technology-based learning', in *Journal of Further and Higher Education*, 26(1), 29–42.

CIRC, 2001, *Construct for Excellence*, Hong Kong, Construction Industry Review Committee.

Clarke, T. and Hermens, A. 'Corporate developments and strategic alliances in e-learning', in *Education and Training*, 43(4), 256–267.

Cox, A. and Townsend, M., 1998, *Strategic Procurement in Construction*, London, Thomas Telford Publishing.

Egan, J., 1998, *Rethinking Construction*, London, Department of the Environment, Transport and the Regions, London, HMSO.

Fry, K., 2001, 'E-learning markets and providers: some issues and prospects', in *Education and Training*, 43(4/5), 233–239.

Gibb, A.A., 1009, 'Small firms training and competitiveness: building upon the small business as a learning organization', in *International Small Business Journal*, 15(3), 13–29.

Hall, G., 1991, *Non-financial Factors Associated with Insolvency amongst Small Firms in Construction*, Working Paper No. 214, Manchester, Manchester Business School.

Hall, G., 1995, *Surviving and Prospering in the Small Firm Sector*, London, Routledge.

Hammond, M., 2000, 'Communication within on-line forums: the opportunities, the constraints and the value of a communicative approach', in *Computers and Education*, 35(4), 251–262.

Hiltz, S.R., Coppola, N., Rotter, N., Turoff, M. and Benbaum-Fich, R., 2000, 'Measuring the importance of collaborative learning for the effectiveness of ALN: A multi-measure, multi-method approach', in *ALN Journal*, 5(2), http://www.aln.org/publications/jaln/v4n2_hiltz.asp (2000), (accessed June 2005).

Hughes, S.C., Wickersham, L., Ryan-Jones, D.L. and Smith, S.A., 2002, 'Overcoming social and psychological barriers to effective on-line collaboration', in *Educational Technology and Society*, 5(1), 86–92.

Jones, E. and Martinez, M., 2001, 'Learning orientations in university web-based courses', in *W. Lawrence-Fowler and A. Hasebrook (eds) Proceedings of Webnet 2001, World Conference on the WWW and Internet, Orlando, Florida, October 23–27;* Florida, AACE, 621–626.

Kaplan, J., 1996, 'Small business is big business', in *ASCE Journal of Practice Periodical on Structural Design and Construction*, 1(3), 78.

Keeble, D. and Wilkinson, F., 1999, 'Collective learning and knowledge development in the evolution of regional clusters of high technology SMEs in Europe', in Regional Studies, 33(4), 295–303.

Latham, M., 1994, *Constructing the Team*, London, HMSO.

Masslen, G., 2004, 'Bricks and mortar still beat Web for learning', *in South China Morning Post*, Education Section, 27 November, Hong Kong, page E2.

Matthews, J., Thorpe, A. and Tyler, A.,1997, 'A comparative study of subcontracting in Hong Kong', in *Campus Construction Papers*, May, Ascot, UK, CIOB, 13–16.

Miller, C., Packham, G. and Thomas, B., 2002, 'Harmonization between main contractors and subcontractors: a pre-requisite for lean construction?', in *Journal of Construction Research*, 3(1), 67–82.

Morton, R. and Jagger, D., 1995, *Design and the Economics of Building*, London, E & FN Spon.

Packham, G.A. and Miller, C.J.M., 2000, 'Peer-assisted student support: a new approach to learning', in *Journal of Further and Higher Education*, 24(1), 55–65.

Raelin, J., 2001, *Work Based Learning: The New Frontier of Management Development*, London, Prentice Hall.

Rahman, M.M., Kumaraswamy, M.M., Rowlinson, S. and Sze, E., 2003, 'Performance improvements through flexible organizational cultures', in R. Fellows and A. Liu (eds) *Proceedings of the CIB TG23 International Conference, The University of Hong Kong, Hong Kong*, 26–27 October, CD Rom.

Rahman, M.M., Kumaraswamy, M.M., Ng, S.T., Palaneeswaran, E., Lam, E. and Ugwu, O.O., 2004a, 'Integrating SMCs into construction value chains', in P. Poh, A. Mackenzie and C. Katsanis (eds) *Proceedings of the 1st International Conference of World of Construction Project Management (WCPM2004), Ryerson University, Toronto, May 2004,* Toronto, Ryerson University 552–563.

Rahman, M.M., Kumaraswamy, M.M., Ng, T., Lam, E., Ho, E., Lee, S., Palaneeswaran, E. and Ugwu, O.O., 2004b, 'Needs of SMCs in Hong Kong: information management and IT', in C.J. Katsins *et al* (eds), *Proceedings of the Annual Meeting of CIB W102 Information and Knowledge Management in Building, Ryerson University, Toronto, April 2004*, Toronto, CIB 59–66.

Ravet, S. and Layte, M., 2001, *Technology-Based Training*, London, Kogan Page.

Stevens, G., 1999, 'The role of logistics and IT in the European enterprise' in E. Hadjiconstantinou (ed.), *Quick Response in the Supply Chain*, New York, Springer-Verlag, 11–20.

Stewart, R., Miller, C., Mohamed, S. and Packham, G., 2003, 'Sustainable development of small construction enterprises: IT impediments focus', in *WEI Working Papers Series 32* (2003), 1–15.

Thorne, K. and Mackay, D., 2001, *Everything You Ever Needed to Know About Training*, London, Kogan Page.

Ugwu, O.O., Kumaraswamy, M.M., Rahman, M.M. and Ng, S.T., 2003, 'IT tools for collaborative working in relationally integrated supply chains', in F. Bontempi (ed.) *Proceedings of the 2nd International Structural Engineering and Construction Conference (ISEC-02), Rome, Italy, September 2003, 1*, Lisse, The Netherlands, A. A. Balkema, 217–222.

Wall, J. and Ahmed, V., 2004, 'E-learning and pedagogical challenges in construction management: bridging the gap between academia and industry', in F. Khosrowshahi (ed.) *Proceedings of the 20th Annual Conference of ARCOM, Edinburgh, UK, September 2004*, Reading, ARCOM, 603–612.

ARTICLE

Retrofitting E-learning to an Existing Distance Learning Course

A Case Study

Stuart Allan, Keith Jones and Simon Walker

Abstract

The School of Architecture and Construction of the University of Greenwich has successfully delivered a range of distance learning programmes, mainly in Hong Kong, Malaysia and China, since the early 1990s. The method of delivery has been through paper-based study material and study weekends held twice a year at each centre. Recently, the distance learning market has become more fragmented, diverse and scattered with students expecting multimedia programme delivery. In 2003, the decision was taken to examine the potential of electronic teaching and learning systems to meet changing student expectations. This paper describes the experiences of the authors as they retrofitted an e-learning solution to an existing distance learning course. The paper, through the use of a case study methodology, critically reviews the various stages of the retrofitting process and highlights the key factors that affected its progress. The paper concludes that, while it was technically feasible to retrofit an e-learning solution to a traditional paper-based course, the trade-offs between the potential benefits and associated costs, and the need to clearly articulate these to other stakeholders in the distance learning programme, was a major inhibitor on both the scale and speed of development.

■ *Keywords* – e-learning; blended learning; virtual learning environment; curriculum development

BACKGROUND TO THE DISTANCE LEARNING PROGRAMMES

The School of Architecture and Construction of the University of Greenwich has provided a range of distance learning programmes since the early 1990s. The first to be offered were built environment programmes and these were followed in the late 1990s by occupational health and safety programmes. The programmes were initially delivered through centres in Malaysia, Hong Kong, Australia and the UK. Students were registered to specific centres and expected to attend a number of study schools each year. New centres in China and Trinidad have been established or are in the process of being established.

In addition to the study schools, students were provided with a comprehensive study pack containing a study guide, course guide and a selection of two or three textbooks. The study guide, which formed the backbone of the course, was a fairly thick A4 ring-bound information source intended to deliver similar material to that which a student would receive during lectures on a taught programme. The course guide contained administrative information, including assignments and was intended to guide the student through the course. The textbooks were selected to support the whole learning process. These were largely paper-based although some material was transferred on to read-only compact discs.

Over the past few years, the school's distance learning market has become more fragmented, diverse and scattered. Although the above courses were still delivered to students attending study weekends, increasing numbers of students were located in undeveloped parts of the world and were geographically isolated from the existing centres. In these circumstances, students were registered in the UK and invited to attend the study schools held at the University of Greenwich. Many, however, found this difficult and expensive, preferring instead to study with other providers who could more easily support them at their location. Thus, while the school had a successful record of providing distance learning, it found its market share reducing in the face of increased competition, especially from new providers who had developed materials in an electronic format which included online mediation and support. In comparison with these solutions, the school's traditional paper-based approach looked somewhat dated and in 2003 the school recognized the need to review its courses in the light of e-learning developments. However, in establishing the review, the school management were insistent that whatever electronic solution was adopted had to protect the significant investment, in terms of time, experience and money tied up in the 'traditional' paper-based system. In essence, the authors were posed the problem of developing an e-learning solution that could be retrospectively fitted to the existing paper-based course in a way that protected the investment while maximising the attractiveness to remote (to the existing centres) students. This paper describes the process undertaken by the authors.

The paper is based on a case study methodology in which the actions of the authors have been critically reviewed in the light of the experiences gained. The paper does not seek to critically review alternative pedagogic approaches to e-learning (although it does provide some general background to the blended approach to place the project in context), nor does it argue for new tools for the delivery of e-learning. Instead, the case study describes the practical steps that the authors took when faced with the challenge of developing a real course for real students in real time. From these experiences, the authors have drawn general conclusions that they believe will assist other built environment educators facing similar challenges.

A BLENDED APPROACH TO E-LEARNING

In the UK, within the further and higher education sectors, a number of Government initiatives – *The Dearing Report* (Dearing, 1997), *The Higginson Report* (FEFC, 1996), *Success for All* (DfES, 2002), *Get on with IT* (Morrison, 2002), *Managing Inspection and ILT* (Powell *et al*, 2003), *The Future of Higher Education* (DfES, 2003a) and *Towards a Unified e-learning strategy (consultation)* (DfES, 2003b) – identified the need to integrate information and communication technology (ICT) tools into teaching and learning, support the development of innovative information and learning technologies (ILT) and encourage more innovative approaches to teaching and learning to capture the e-learning advantage. New terminology to describe this change has emerged and even the term e-learning, used to describe technology specifically focused on teaching and learning rather than management systems, is itself undergoing change – the 'e' is not so much concerned with 'electronic' as 'enhancement' (JISC, 2004a:10). Further, there is an emerging consensus that learning and teaching can be improved through the selective and appropriate use of technology. In the consultation summary document *Towards a Unified e-learning Strategy*, e-learning is defined as the capacity to improve.

> *The quality of learning through using interactive computers, online communications, and information systems in ways that other teaching methods cannot match. (DfES, 2003b)*

In *Towards a Unified eLearning Strategy*, the Department for Education and Skills (2003b: 51) perceived e-learning as having the potential to provide rich and increasingly differentiated learning opportunities, with individual empowerment and personalization at the heart of the development process. Learning providers should not only develop flexibility over the time, place and mode of study, but make learning resources, peer and tutor group interaction, and universities themselves more accessible.

Early perceptions of e-learning suggested the approach would be a 24/7, just-in-time experience where learners would surf the web, meeting in cyberspace but never needing to gather in person with their peers or tutors. Over the past decade, experience has moderated this vision, leaving in its wake a more realistic picture of the selective, appropriate use of various technologies used in face-to-face, distance or a combination of these modes.

> *'Blended learning' is the key now, I think; something that acknowledges the advantages that ICT can bring in terms of flexibility of delivery and access in the home, the workplace, community centres and so on. I think it's fair to say that we're still feeling our way in this area... Blended forms of learning will be the norm with technology being used in increasingly student-centered environments. It will be interesting to see the effects of this on curriculum development. The greater flexibility of technology will encourage a 'just-for-you' culture within institutions, developing not only bespoke courseware and learning materials, but whole courses of study, delivered into non-traditional surroundings at work or at home. Lifelong learning, continuing personal and professional development will be central. (JISC, 2004b)*

While blended learning is still a relatively new concept, models have emerged from the use of e-learning in a range of delivery, teaching and learning situations. These models are characterized by a number of features including a shift in emphasis from teacher- to student-centred learning, transmission of old knowledge to the construction of new knowledge, behaviouristic to humanistic, inauthentic and content-free to authentic and context-specific tasks.

A very useful description of the different ways that technology can be used to support teaching and learning was developed by Scribbins and Powell (2003). The e-learning fan (Figure 1) illustrates the different contexts within which learners and teachers operate. The seven segments represent the different ways that e-learning is used, from providing support to traditional face-to-face teaching to the ultimate e-learning experience, i.e. pure remote learning where all materials and tutoring takes place online. It should be noted that

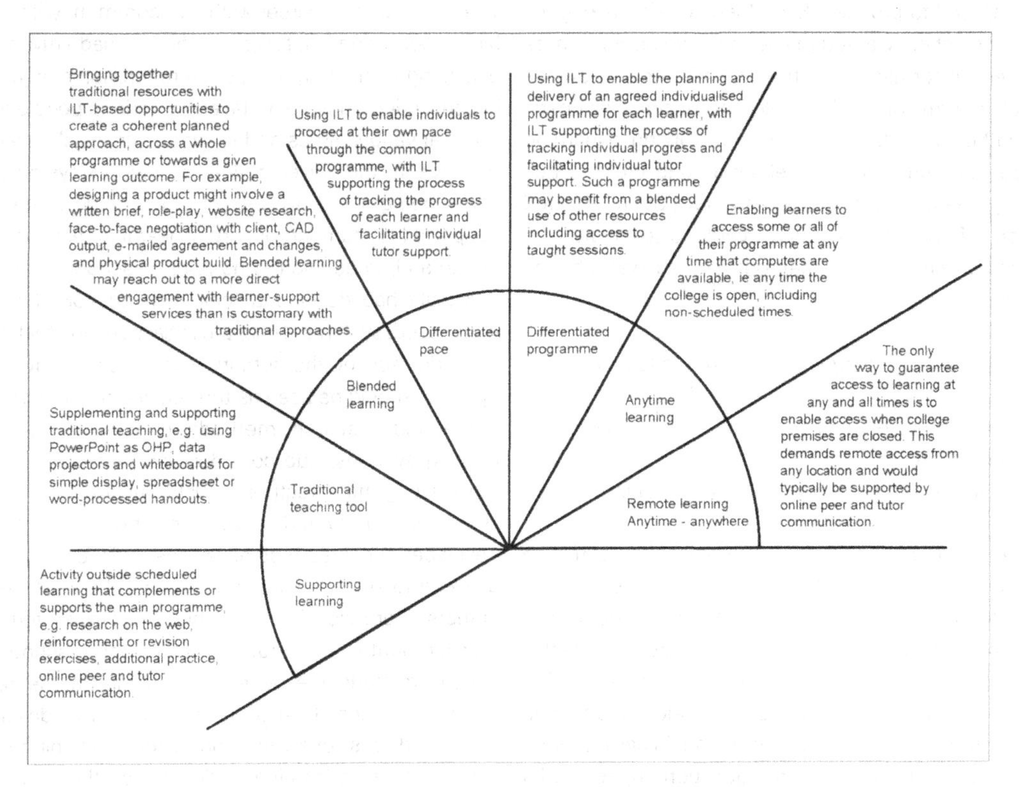

FIGURE 1 The e-learning fan (as cited by Powell *et al*, 2003)

Scribbins and Powell were keen to point out that no one context is valued above another, but e-learning has a place in all these contexts to a greater or lesser extent.

A BLENDED APPROACH TO DISTANCE LEARNING

From a consideration of the e-learning fan and the needs of this project, where students were studying at a distance, in remote locations and where a large investment was tied up in existing systems, the blended approach was the one that appeared to the authors to offer the best solution. However, before the development of a blended approach to distance learning could begin, the support of the rest of the distance learning team was required and, even though everyone acknowledged the need to do something, many were sceptical of adopting the blended approach. In essence they believed that the costs involved could not justify the potential benefits. Indeed, it was not until a funding opportunity presented itself that the school finally agreed to proceed. Even then certain colleagues within the school wanted to use the development more as a vehicle for assessing the potential impact on the school (in terms of training, skills needs, financial costs, etc.) rather than developing a fully working e-learning course. After considerable debate among the distance learning team, it was agreed to develop a blended version of one course as a pilot to evaluate the approach. The research methods course was chosen for the pilot as it:

- encompassed a range of teaching and learning activities
- was a compulsory course for all distance learning students, and
- two of the authors were part of the teaching team.

The research methods course required students to engage in a series of tasks that culminated in the presentation of a research proposal which could form the basis of their dissertation. To undertake these tasks, students were required to engage in brainstorming activities to identify research issues of relevance to their programme of study, to undertake an initial literature review, to formulate a research question supported by specific aims, objectives and, where appropriate, a hypothesis, to outline an appropriate research methodology to address the aims and objectives and to produce a work plan to show how the project would progress. In retrofitting a blended e-learning solution to this course, the authors had to consider:

- what technology existed to support the 'blended' learning approach?
- what changes would be required to the research methods course?
- what should be the nature of the blend?
- what would the benefits be to the school and university and, more importantly, to the students?

To support the above, the authors developed a series of toolkits which, if successful, could be applied to the development of other blended distance learning courses.

TECHNOLOGY TO SUPPORT THE BLENDED APPROACH

The University of Greenwich, in common with many higher education institutions in the UK, had arrived at its e-learning technology infrastructure through a process of historical decisions being taken by various people in key posts at various times and in various financial situations. As a result, the university had a range of systems that were being used to deliver e-learning. It centrally supported a number of virtual learning environments (VLEs) and, in addition, a number of schools within the university had developed their own bespoke (primarily web-based) systems. Thus, the initial question that had to be addressed by the authors was which, if any, of the systems should be used as the vehicle for the delivery of a blended research methods course. In reviewing existing systems, due consideration was given to both the existing infrastructure available at the University of Greenwich (it would have been pointless to have developed a course that lacked the logistical support to deliver it to students) and the long-term benefits that the authors were seeking to achieve (both in terms of developmental infrastructure within the school and enhanced student experiences). To inform the review process, a series of performance criteria were developed (through discussions with other distance learning tutors, existing distance learning students and other colleagues within the university). The VLE:

- had to be accessible by anybody, anywhere, at anytime
- should not require the user to purchase third-party software to access it
- from a learner's perspective should accommodate the most popular commercial software (e.g. Windows-based products)
- should be simple to use
- should be able to accommodate the range of document formats that the school regularly used (text, numeric and graphic formats)
- should be simple to use by academics who were not IT experts
- should be supported by the university.

From the review, WebCT Campus Edition was ultimately selected as the VLE for this project because it satisfied all of the above and additionally:

- it provided an effective repository for content
- was accessible and compliant to World Wide Web Consortium (W3C) standards
- had an extensive toolkit that was relatively easy to use
- it already interfaced with other university-wide systems and, as such, could manage student logon details as part of the student registration process thus minimising administration.

There were some disadvantages identified during the evaluation process, including the need to use hyper text mark-up language (HTML) to programme pages, unsophisticated assessment tools, a didactic pedagogical approach, and the time taken to develop suitable materials. However, similar criticisms could be levelled at the other alternatives evaluated. The final and probably critical factor that influenced the decision to select WebCT as the VLE for this project was the support offered by the university. WebCT was its preferred VLE system and the university had put in place the infrastructure to run WebCT training and provided technical support for the transfer of information on to the WebCT platform. In the end, this degree of certainty made using WebCT more attractive than other alternatives.

REVIEWING THE RESEARCH METHODS COURSE

The next task was to critically examine the delivery of the research methods course to identify what enhancements could be brought about by a blended approach. However, this had to be done alongside an understanding of the opportunities and limitations that existed through the tools available in WebCT. In particular, to what extent could WebCT be used to improve the delivery of the course and achieve the desired blended learning approach?

The initial analysis of the research methods course used Ryan and Walker's (2003) e-learning approaches grid (Figure 2). The authors adapted the grid to assist them in better understanding the balance between face-to-face and online learning that suited the research methods course. Considering the course against a series of scales (product v. process, deductive v. inductive, content laden v. content free, individual learning v. collaborative learning, closed format v. open format) produced a visual representation of the learning approaches that were available. The analysis of the research methods course was carried out by the authors who had delivered the course for the previous five years. The reasoning behind the positioning of the course against each scale can be summarized as follows.

PRODUCT V. PROCESS

The primary aim of the course was to deliver specific skills that the student would require in order to successfully complete their dissertation. A secondary aim was to provide experience in identifying research questions and proposing alternative methods for their investigation. Based on the relative balance between the two, research methods was positioned towards the 'product' end of the scale.

DEDUCTIVE V. INDUCTIVE AND CONTENT LADEN V. CONTENT FREE

Although the course was fairly tightly structured and content rich (with regular examples used to guide the student through the study material), it used an open negotiated project as the primary learning vehicle and assessment tool. For this reason, the course was positioned mid-way along the 'deductive/inductive' and the 'content-laden/content-free' scales.

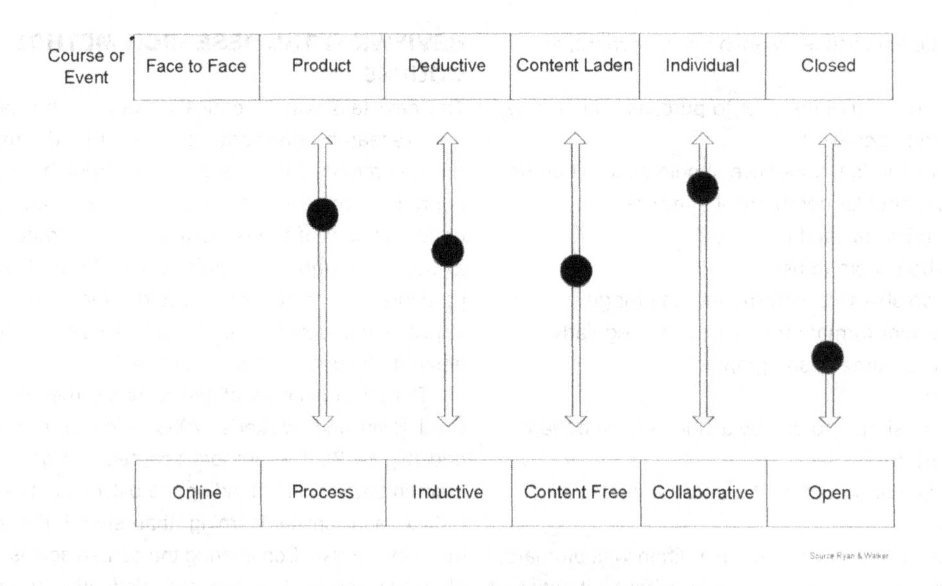

FIGURE 2 Adaptation of Ryan and Walker's e-learning approaches grid

INDIVIDUAL V. COLLABORATIVE

Because of the personalized nature of the negotiated project, the research methods course viewed the student as an individual learner who needed to work closely with their tutor rather than in collaboration with students. The latter was encouraged where students were feeling isolated to provide a level of mutual support and, in some cases, peer-to-peer review. As such, the research methods course was positioned very close to the 'individual' end of this scale.

CLOSED FORMAT V. OPEN FORMAT

Although there was a suggested progression route through the study material, the course did not require students to engage in a specific activity each week nor did it require information to be passed from one student to another. Thus the student had a fair degree of flexibility as to exactly when they studied and the order in which they tackled the various tasks set. As such, the research methods course was positioned towards the 'open' end of the scale.

While the e-learning approaches grid provided a visual expression of the potential balance between online and face-to-face delivery and reinforced the view that research methods would indeed lend itself to a blended approach, it did not really provide a toolkit that focused on the specific developments that would be needed to retrospectively fit a VLE solution to the existing course. To do this, a more detailed tool was needed that could identify specific opportunities for development and prioritize these in terms of an enhanced student experience. From the experience of the authors delivering the research methods course, a matrix cross-referencing the delivery of the course with the tools available through WebCT was developed. The matrix is shown in Table 1.

The matrix was used in two ways. First, an initial view of the importance of each of the WebCT tools in relation to the various aspects of the teaching process was assessed using a three-star rating system (*** very important, ** neutral importance, * not important). This analysis was based on an initial 'gut' feeling of the tutors towards the various tools and allowed an immediate view to be taken as to their usefulness. Following this review, a more detailed assessment of the tools was performed by considering how the tool could be used rather than just expressing the view that it should be useful. In this analysis, a numeric scoring system (1 'not useful' – 20 'very useful') was used for each tool. Again the scoring was undertaken independently by the

TABLE 1 Teaching process/VLE evaluation matrix

	TEACHING PROCESS																				
	TEACHING					COMMUNICATION					ASSESSMENT						ADMINISTRATION				
WEBCT TOOLS	LECTURES	SEMINARS	TUTORIALS	ONE-TO-ONE	PRIVATE STUDY	LARGE GROUPS	SMALL GROUPS	ONE-TO-ONE	SYNCHRONOUS	ASYNCHRONOUS	GROUP	INDIVIDUAL	FORMAL	INFORMAL	FORMATIVE	SUMMATIVE	STUDENT	LECTURER	SCHOOL	UNIVERSITY	SCORE
Course content																					
Syllabus																					
Content module																					
Glossary																					
Image database																					
Index																					
Content utility																					
Search																					
Compile																					
Resume course																					
CD-Rom																					
Communication tools																					
Discussions																					
Mail																					
Chat																					
Whiteboard																					
Calendar																					
Student tips																					
Evaluation & activity																					
Quiz																					
Survey																					
Assignments																					
Student presentation																					
Student homepages																					
Student tools																					
My progress																					
My grades																					
Language selector																					

authors. The average scores awarded to each WebCT tool are shown in Table 2.

In many respects, the scoring exercise confirmed what had been anticipated, with the communication and assignment tools both scoring highly. What was perhaps less expected was the high scoring of the calendar tool but, interestingly, as the course developed, its usefulness became more apparent. There was some surprise at the high score achieved by the search tool and the authors have yet to discover why that scored so highly. The scores awarded to the syllabus and content module were to be expected although these were not envisaged as the main mechanisms for achieving a blended learning approach.

TABLE 2 Result of the teaching process/VLE evaluation process

WEBCT TOOL	SCORE
Syllabus	12
Content module	15
Glossary	5
Image database	4
Index	7
Search	14
Compile	2
Resume course	4
CD-Rom	4
Discussions	12
Mail	9
Chat	10
Whiteboard	5
Calendar	14
Student tips	9
Quiz	4
Survey	6
Assignments	16
Student presentation	5
Student homepages	3
My progress	11
My grades	12
Language selector	1

THE RESEARCH METHODS WEBCT COURSE

The final process that was required was the application/integration of the various WebCT tools into the existing research methods course. To achieve this, the authors considered how each of the WebCT tools could be used with the existing distance learning material. The criteria used for this part of the exercise were to enhance the student experience while, at the same time, optimizing tutors' time.

From the analysis of the pedagogic mix of the research methods course (Figure 2), it was apparent that the most suitable activities for online delivery were those associated with the open format aspects of the course. While there was no prescribed route through the study material, experience of the authors had shown that students did benefit from some structure to their learning. Although different students started their learning at different points along the knowledge continuum, most of them experienced similar problems as their learning progressed. These problems were usually associated with aspects of assessment. Thus it was this aspect of the course that the online solution was developed to tackle. The blended mix for the course is shown in Figure 3.

From Figure 3 it can be seen that the assessment tasks, rather than the study guide, now formed the backbone of the learning structure. The challenge to the authors was to develop the blended infrastructure to support this approach. The solution adopted was to support each stage of the student learning through a mixture of the traditional (face-to-face activities as part of the study weekend and textual support through the study guide and key supporting texts) and a new electronic approach (PowerPoint presentations, group discussions, one-to-one discussions, online library support, online submission and online feedback).

Initially, online developments focused on the use of the communication, calendar and assignment tools although the course outline, course information and study guides were also uploaded as Word files. The first problems arose when attempts were made to uplift PowerPoint presentations into the WebCT environment. Although WebCT was WC3 compliant and, as such, PowerPoint should have been accessible, the authors experienced significant problems that delayed developments and resulted in implementation being put back while alternative solutions were sought. Eventually a decision was taken to replace all the PowerPoint presentations (which were part of the original face-to-face aspect of the research methods course) with HTML pages. Although this provided an enhanced system for the student, it did require a significant investment to effectively recreate something that already existed. In terms of cost–benefit considerations, this was a major negative issue and provided 'ammunition' to those within the school who believed e-learning was too expensive an activity.

To support the discussion forums envisaged as part of the blended approach, it was necessary to develop both synchronous and asynchronous communication channels. However, it quickly became apparent that there would be significant logistical problems in utilizing the synchronous tool, where staff and students live in different time zones. That is not to say that the

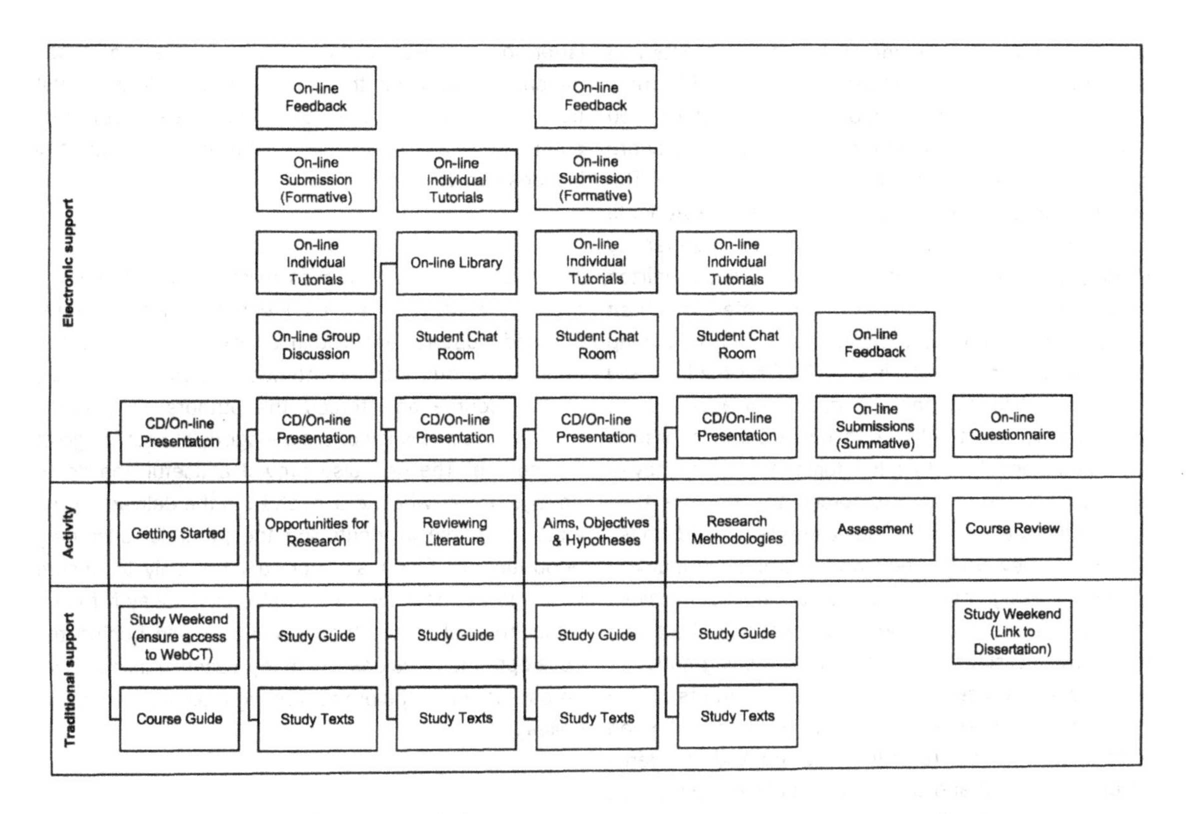

FIGURE 3 A blended approach to research methods

synchronous chat communication was dismissed as a tool for peer-to-peer communication, but it was not enacted as a formal means of communication between the student and tutor.

With regards to the development of the asynchronous channel, what had become apparent to the authors through their experience teaching the course was that the immediacy of e-mail produced in the student the expectation that their queries would be answered by return, if not within the hour at least within the day. This had put considerable pressure on tutors and had resulted in ad-hoc systems being established to manage student expectations. By way of an anecdotal story, one tutor e-mailed all students explaining that all non-urgent e-mails would only be addressed at a given time each week. Not surprisingly, the tutor then only received e-mails that were marked 'urgent', 'very urgent' and 'extremely urgent' and a further e-mail had to be sent explaining that it was the tutor's definition of urgent and not the students' that would determine the response strategy. While one can understand the tutor's response, the failure to manage the students' expectations caused a level of student dissatisfaction that the authors wanted to avoid in the blended approach. To this end, it was decided to utilize the asynchronous communication tool to support a series of managed discussion sessions in which open discussions were encouraged between students and tutors for each aspect of the assignment. These discussions would start on a specific date, run for a specific time and then be archived for future reference by all students on the course. This way, time and location differences became irrelevant and even if students were not ready to engage actively in the discussion, they could benefit from the discourse when it was more relevant to them.

The final aspect of the blended approach was the use of the assignment tool to manage both formative and summative assessment. In the past, students had submitted their assignment tasks in a range of formats using different communication channels (embedded e-mail, document attachments, various software programmes, fax and surface mail, among others). The assignment tool allowed the authors to require students to submit their work in a prescribed manner by downloading a series of templates (which utilized Microsoft Office software), completing their assignments using the templates and then uploading their submissions directly into WebCT for marking and feedback. While this, on the surface, may appear to be over-bureaucratic, it did have many advantages for both the student and the tutors. In addition to consistency of approach, the student was forced to structure their work in line with prescribed guidelines (e.g. word limits), a skill that they would need when completing many of their other assignments, and ensured that submissions were in a format that could be accessed by the tutors (on many occasions the students were using software versions in advance of those available to tutors). From the tutors' perspective, the system allowed submissions to be recorded and tracked (students receive a receipt) and provided a marking regime that was consistent for all students.

The blended approach using WebCT outlined above has recently been implemented with the first cohort of students and early indications suggest that it has been well received.

LESSONS LEARNT DURING THE DEVELOPMENT OF THE COURSE

The development of a blended approach to the delivery of the distance learning research methods course exposed a number of issues that, if not unexpected at the outset, presented the authors with a range of questions that will have to be addressed before a similar approach is rolled out across the rest of the distance learning programme. These lessons fall into two groups, those associated with the blended pedagogy and those related to the VLE tools used by the authors. With respect to the former, the authors believe that a blended approach does offer significant opportunities to enhance traditionally delivered courses through the use of e-learning tools without having to recreate the course from scratch. With respect to the latter, the promises of the VLE toolkits are sometimes difficult to realize and this is a major obstacle that must be overcome before significant developments commence. From this study, the following general lessons can be drawn.

DECIDING WHAT TO DO

The decision-making process followed by the authors provided a sound basis for identification of potential blended opportunities within an existing course. The approaches grid focused attention on various aspects of the course and forced the authors to critically evaluate the relevance of the pedagogy at each stage of the course. The grid also proved a useful vehicle to engage others who were involved in the delivery of the existing course in identifying the potential e-learning opportunities. This was important not only to ensure buy-in to the new system but also in the way it forced tutors to explore the theory and rationale behind the development decisions, in many cases making them ambassadors for the project among their more sceptical colleagues.

DECIDING HOW TO DO IT

Once the general decision had been made to proceed with development of the e-learning course, obtaining a thorough understanding of the opportunities offered by the various toolkits was essential. The use of the evaluation matrix provided the basis not only to identify how WebCT could be used to support the blended approach but also identified those aspects of WebCT that the authors needed to become more familiar with. With the benefit of hindsight, the matrix was probably too complex (requiring detailed knowledge of the specifics of the VLE) to be widely applicable to all developments. The authors are currently reviewing this model with a view to producing a simpler version for use by other colleagues when considering turning their traditional distance learning courses into WebCT-enabled blended courses.

The final toolkit that was used enabled the authors to articulate what it was they were trying to do and how the various components of the blended approach would be integrated together. Again this toolkit proved useful, not only to those directly involved in the development of the course, but also as a vehicle to engage other tutors in the process. Everyone involved with the research methods course, including those associated with its administration as well as the academic tutors, understood the rationale behind the approach and saw how their bit of the solution fitted in. Again this approach is being used by other colleagues as they examine the possible development of blended versions of their courses.

DEVELOPING THE WEBCT COURSE

Although all the background research identified WebCT as a suitable VLE template to develop a blended learning approach of the research methods course, the development process was not straightforward. The course was developed using WebCT Campus Edition 3.8 and a number of teething problems were experienced. Many of these problems were resolved as the authors became more familiar with the software and with the introduction of version 4.1, but it left the development team with concerns over the ability of VLE software to deliver against developers' initial expectations. With the benefit of hindsight, the authors should have been less ambitious in what they attempted and should have invested more time in fundamental WebCT training before the developments started. In this way, both the developers and their colleagues' expectations could have been managed better.

CONCLUDING COMMENTS AND FUTURE DEVELOPMENTS

The aim of this project was to examine whether an e-learning solution could be retrospectively fitted to an existing course in such a way that protected the original investment while providing the student with an enhanced learning experience. Although field trials with students are still ongoing, the authors believe they achieved their aim. However, the procedure that they went through did highlight some issues that still need to be addressed. As the reality of the development process became clear, there was a constant tension between realizing the full benefits that adopting an e-learning approach could bring and optimizing the use of the existing paper-based material. Indeed, in order for the distance learning courses to take full advantage of online learning, most of the existing paper-based material will have to be reviewed and revised in the near future. One question that remains is whether this can realistically be achieved through the routine maintenance associated with the distance learning course or whether the e-learning developments will have to wait until the whole programme is fundamentally reviewed. This question is currently with the school's management team.

AUTHOR CONTACT DETAILS

Stuart Allan and Keith Jones: School of Architecture and Construction, University of Greenwich, Avery Hill Campus, Bexley Road, Eltham, London, SE9 2PQ, UK. Tel: +44 (0) 208 331 8000, fax: +44 (0) 208 331 9105, e-mail: s.allan@gre.ac.uk.

Simon Walker: School of Education and Training, University of Greenwich, Old Royal Naval College, Park Row, Greenwich, London, SE10 9LS, UK. Tel: +44(0) 208 331 9238, fax: +44(0) 208 331 8145, e-mail: s.walker@gre.ac.uk

REFERENCES

Dearing, R., 1997, 'Higher education in the learning society', http://www.leeds.ac.uk/educol/ncihe/ (accessed 11 March 2005).

DfES, 2002, 'Success for all: reforming further education and training – our vision for the future', http://ferl.becta.org.uk/display.cfm?resID=4764 (accessed 15 March 2005).

DfES, 2003a, 'The future of higher education', http://www.dfes.gov.uk/consultations (accessed 13 March 2005).

DfES, 2003b, 'Towards a unified e-learning strategy', http://www.dfes.gov.uk/consultations (accessed 13 March 2005).

FEFC, 1996, *Report of the FEFC Learning and Technology Committee* (Higginson Report), currently out of print.

JISC, 2004a, 'Effective practise with e-learning', HEFCE, http://www.jisc.ac.uk/uploaded_documents/ACF5DO.pdf (accessed 21 March 2005).

JISC, 2004b, 'Looking to the future', interview with Chief Executive Sir Howard Newby, http://www.jisc.ac.uk/interview (accessed 21 March 2004).

Morrison, S. (chairman), 2002, *Get on with IT – The Post-16 E-Learning Strategy*, Task Force Report, http://ferl.becta.org.uk/content_files/pages/keydocs_current/gov/GetOnwithIT_Doc.pdf (accessed 12 March 2005).

Powell, B., Knight, S. and Smith, R., 2003, 'Managing inspection and ILT', , BECTA, http://ferl.becta.org.uk/content_files/ferl/pages/focus_areas/ilt_inspection/raising_%20standard_%20initiative/Managing%20ILT.pdf (accessed 21 March 2005).

Ryan, M. and Walker, S., 2003, 'Experimenting with and analysing approaches to e-learning within a staff development context', in *ED-MEDIA & ED-TELECOM World Conference (AACE), Hawaii*, 23–28 June, 2003.

ARTICLE

Developing a Knowledge Centric Approach to Construction Education

David Boyd

Abstract

Knowledge management has been an in-vogue management initiative for a number of years but the obvious link to education has not been made except for the use of Web-delivered information. This paper establishes a knowledge centric approach to construction education and reports on the application of this to a master's conversion course. The knowledge centric approach develops a capability to work with incomplete knowledge. Incomplete knowledge is a growing problem in the industry as projects and technology become more complex; for conversion graduates it is the critical factor. The knowledge centric idea was based on research into the knowledge of competent practitioners. This highlighted the growing tension between the need for a deeper understanding of construction activity as solution composition, as against the application of technical knowledge. The approach requires abilities – to appreciate self-knowledge, to access different knowledge (technical, organizational and human relational) in situations, to understand the social operation of knowledge, to negotiate action under competing knowledge and to develop new knowledge. The challenge of the knowledge centric approach is to create practitioners who can act effectively with incomplete knowledge.

■ *Keywords* – Conversion education; knowledge management; incomplete knowledge

INTRODUCTION

This paper comes out of a practical concern for the inadequacies of construction education and also from the need for new and better practitioners to be created rapidly. Construction courses have been in existence for many years. These have been substantively of a technical nature. There was a belief in the 1960s that an understanding of the science behind practice would enable practitioners to perceive, devise solutions and act in practice more effectively (Franklin, 1992). This idea meant that research into practice involved finding out how the parts worked in more and more detail. This information would then be transferred to student practitioners in formal lectures or texts by people who were investigating these parts. Such thinking is still widespread and is the constitution of many construction courses. However, there have been many dissenters. Franklin (1992) states with a sense of warning that: 'Consequently, somebody can today go to a university and learn how to build bridges from someone who has never built a bridge.'

Indeed, practice to some extent believes that academic education is problematic to it (Business Round Table, 1982; Anderson, 1992; Andrews and Derbyshire, 1993; Strategic Forum, 2003; Lenard, 2004); even some of the industry's wider problems have been blamed on education (Latham, 1994). Educators are often told that practice is more than can be taught or learnt at college. In the US, this has caused the American Council for Construction Education to rewrite its curriculum (ACCE, 1995). In the UK, it has given rise to new concepts of practice education based on testable competencies with National Vocational Qualifications (NVQs) and now national benchmarks, which attempt to prescribe the nature of education.

Thus, there is a continuing tension between education and work, between the deeper understanding of activity and the ability to do the job.

This tension has been magnified by the desire to change practice in the industry where techniques of management of organizations and people are becoming important attributes (CIB, 1996). The developments from Latham (1994) and Egan (1998) have criticized the silo attitude of professions, the poor attention to clients and the destructive relationships of procurement. The paradox is that the industry says novice practitioners learn best in practice; however, what practitioners pick up (to some extent) in practice is negative. Change requires the industry to not just do things differently, but also to think differently, otherwise older practices become the only solutions available when problems are imminent. Education needs to change to stop tacitly supporting these negative practices.

Undergraduate construction courses have been failing to recruit entrants of the quantity and quality that are required by the industry (Gann and Salter, 1999; CEBE, 2003; Strategic Forum, 2003). This is producing a skills gap in the industry, certainly now but also, as importantly, for the future. This is holding back not only the construction industry but also regional and national redevelopment and business. At the same time, there are unemployed graduates from non construction disciplines, because of the mass expansion of higher education and the provision of a wide choice of courses, who would now value the security of a professional career in the construction industry. Thus, the human resource gap can be matched by this source, if the industry can be made attractive and if suitable conversion courses can be designed that meet the needs of both the industry and graduates.

An approach to the rapid conversion of individuals to become effective construction professionals is required. The delivery of undergraduate technical education in a shorter period of time would be one method. However, this may not be physically possible or even desirable for the future careers of these people, given the foregoing negative view of education. A more rewarding approach is to consider the nature of the industry and the nature of professional jobs undertaken, so as to develop individuals to be able to perform and develop better and faster within it. An approach through knowledge management has been taken. This also acknowledges the graduate character of the individuals as well as establishing a postgraduate level of educational content which makes the course more attractive and meaningful to participants. In addition, it takes on board the need to change the industry and the way that it thinks.

Knowledge management (KM) is a drawing together of concepts, techniques and technologies around the perception of the world as information and knowledge (BSI, 2001). In the past we have focused on what organizations did. We lived in a practical world of action. In recent years, through desires for extra efficiency and from the application of computers and indeed from an expanding global view, we have started seeing the world as information and communications (Castells, 1996). KM involves 'the way in which organizations create, find, use, share and organize knowledge. The purpose of KM is to improve performance by making sure people can access and apply the right knowledge, at the right time and the right place' (Payne and Sheehan, 2003). In this, the knowledge itself is the asset rather than a physical product or space or indeed specific skills. It is the separation of knowledge from its context that makes it a tradable asset (Empson, 2001). Many people do not believe in this strong KM model; however, in its focus and exploration there have been significant changes in the way we view the world. In particular, it has forced us to challenge what we know, how we know this and how we work with knowledge (Weick, 1995; Baumard, 1999; Dawson, 2000).

Education has for many years inquired into knowledge and learning, seeing opportunities in changing its perspectives; however, this has almost exclusively been about the media of education (e.g. Laurillard, 2001). There seems to have been a missed opportunity of using KM to challenge what we know, how we acquire knowledge and how we use knowledge in education. Indeed, this paper takes this as its starting point, adding that it is knowledge in practice (Argyris and Schön, 1974) that is critical in construction.

Working with practitioners' perceptions and interpreting from them means that the paper is adopting a phenomenological position (Easterby-Smith *et al*, 1991. This process involves the generation of theories that form different views of the same phenomena. This is a social constructivist view of the world (Easterby-Smith *et al*, 1991) and so objectivity and value neutrality are unavailable in education, as in practice. In this, the paper focuses on trying to understand what is happening in a study of the totality of the conjoint situation in which students are educated whilst working in practice. Theories and concepts act as aids to the interrogation of their experiences, and also the checking of their logical consistency, so that an evaluation of the quality of experience can be undertaken.

PRACTICE KNOWLEDGE

We are calling this a knowledge centric approach adopting ideas and techniques from both knowledge management and personal learning. Knowledge is regarded as an information base plus skills, competencies and capabilities that enable effective practice. The term knowledge centric comes from the conceptions of a maturing knowledge-based organization as it travels from chaotic to centric (Payne and Sheehan, 2004) – see Table 1.

This gives us our challenge in education for practice: first, to identify how knowledge is used and shared; second, to present procedures and tools to assist knowledge use and development; third, to work through the endemic technical, organizational and cultural issues of the industry and education; and finally, to make central the education and practice mission as knowledge for practice. We want students, in the end, to be responsible for their own knowledge management and to enable their companies to develop. The latter of these meets our agenda of changing the industry. Our KM approach to the rapid conversion of individuals to become effective professionals is to consider the nature of the industry, the nature of professional jobs undertaken and the nature of the individuals we will be educating with an aim to develop these individuals to be able to perform and develop better within the industry.

TABLE 1 Knowledge management maturity (Payne and Sheehan, 2004)

Knowledge chaotic	The organization doesn't have many processes and systems for sharing knowledge and people are reluctant to share what they know
Knowledge aware	The organization understands the importance of managing knowledge and has started to identify how knowledge is used and shared, but awareness and understanding of the issues vary across the firm
Knowledge enabled	KM is beginning to benefit your company: procedures and tools are available, but there are still some problems
Knowledge managed	The framework and tools for KM are well established, technical and cultural problems have been solved and there is a KM strategy in place that is updated regularly
Knowledge centric	KM is central to the company mission and its value is measured and reported on

Knowledge management can help us to challenge the norms of our educational delivery, thus allowing development of the subject and its teaching and learning. It allows us to focus on the knowledge that is required to practice. Argyris and Schön (1974) have identified a distinction between practice knowledge and academic knowledge. Knowledge management makes the distinction between explicit knowledge and tacit knowledge (Nonaka and Takeuchi, 1995) which reflects this separation. Macdonald (1995) indicates that professional skills utilize both abstract knowledge (Abbott, 1988) and judgement (Larson, 1977). However, it is the standardization of skills that has allowed the identity, hence definition, of a particular professional service. Most professions have a recognized base of complex, systematic, codified and generalized knowledge, i.e. explicit knowledge, that they bring to bear on activities and which also forms the basis of academic education. This explicit knowledge has been generalized, then analysed into categories that have been researched in detail and this then encapsulated into texts (MacDonald, 1995). However, the exercise of professional judgement is not amenable to standardization or codification. This 'indeterminacy' (Jamous and Peloille, 1970) is significant for all professions, but a balance is struck between codified and indeterminate knowledge for the operation of both the task and professional project (Macdonald, 1995).

Education needs to see practice in a new way including, but not exclusively, the implementation of explicit knowledge (Boyd and Pierce, 2001). The following ideas conceive practitioners' work as an

TABLE 2 Constitution of a practitioner

	RATIONAL	NON-RATIONAL
Outward focus	Knowledge	Tacit knowledge
	Professional and academic theory	Skills, relationships
	Expert power	Intuition, empathy and physical power
Inward focus	Beliefs and values	Fears and compulsions
	Emotional power	Personal drive

engagement with a complex, uncertain and conflicting world that can be synthesized into the dimensions of the model shown in Table 2.

Practitioner knowledge is both rational, i.e. that which can be made explicit, and non-rational, i.e. that which is implicit and cannot be described in words and may not be knowable. Action involves the working with this knowledge both on the outward task and with one's self i.e. inward focus. Thus, the practitioner's personality affects the knowledge they see as important and how it is used both around what they choose to reveal, such as their values, and what lies hidden, such as their drive to succeed (Baron and Byrne, 1994). Patel *et al* (1999) note that the key element of the expert's superior performance lays in the interconnectedness of these knowledge dimensions. Excellence in one dimension alone is not sufficient for superior performance. It is the total individual and his/her embedded skills that make the difference. Education in the past has been particularly good and focused on the rational outward world which yields to being encapsulated and codified. It has only recently explicitly worked in the other three areas.

The individual is extremely important in professional action. One key idea is that implicit knowledge is within the individual. This cannot be expressed, it can only be known to the person who holds it (Polanyi, 1966). Although Wenger (1998) and Baumard (1999) see a social dimension to knowledge, this is exhibited by an individual in relation to a community – in this case, a profession and an industry. Going deeper into professional action involves more indeterminate knowledge, such as intuition, which allows the practitioner to integrate isolated bits of data and experiences into an action picture; thus, practice is a holistic perception of reality that transcends rational ways of knowing (Khatri and Ng, 2000). The process of intuition is very quick and almost automatic and so often can short-circuit stepwise decision making, thus allowing an individual to know almost instantly what the best course of action is (Khatri and Ng, 2000).

The individual in professional action also embodies a set of values although these are often declared as a code of ethics (MacDonald, 1995). However, other attributes come into play at this individual level including emotional intelligence (Goleman, 1996) and an ability to work altruistically and fairly at times against the practitioner's own interest (Pearson, 2003). Such knowledge is not generalizable like abstract knowledge. It has value only in the particular context of practice and consists not of knowledge that is theoretical, but rather is knowledge of 'how to do things' and 'what is the right decision in this context' (Horvath *et al*, 1999). By definition, professionals operate in a dynamic, human world, with all the difficulty that implies, against which they develop skills of coping.

Although we can create models of practice, as in Table 2, our problem is that we have very little detail of how practitioners think in practice. Most research is on the idealized models of practice based on developing further rational codified processes. Practitioners themselves often expound colloquially about their practice and will populate their teaching with anecdotes. These anecdotes are seldom reviewed critically for the effective development of novices and can often be seen as the reproduction of old practice norms. Many of these can be divisive in an industry seeking to change its relationships and its norms of working. Thus, to develop further, requires more work into how practitioners think in practice.

KM RESEARCH INTO PRACTITIONER KNOWLEDGE

During the development of this course, a knowledge management research project was being undertaken which was based on capturing, transforming and disseminating knowledge from the day-to-day practice of project managers and site managers (Boyd *et al*, 2004). This involved these practitioners recording on a Dictaphone each week their handling of a problem event, being debriefed after a series of these events to determine their problem-solving strategies and the wider learning, and company workshops to broadcast the learning. Debriefing (Pearson and Smith, 1985) enables the exploration of the gap between theories in use and espoused theories (Argyris and Schön, 1974) thus allowing practitioners to critically reflect on their practice (Schön, 1983). More than 300 problem-solving events have been debriefed and analysed and this has provided learning at a number of levels, some of which has been useful in the creation and operation of this course. It was realized that this research into practice could also provide results on how expert practitioners think in practice which was exactly what was needed to re-conceive the education of novice practitioners. The results revealed both positive aspects of thinking that needed to be transferred to novice practitioners and also gaps in thinking that the novice practitioners needed to develop strategies to overcome.

The results revealed that practitioners had very sophisticated approaches to problems that were highly individualistic and context dependent. Their thinking was based on past events rather than the normal category-based academic structure of technology, economics, law, business management, etc. Expert practitioners start from a solution base rather than a category base. Successful new solutions are a composite of past events rather than a one-to-one solution to a problem and it is the way the new solution is composed and operated in the particular context and for the particular individual manager that is important. New knowledge is generated through this composition from events and then held within the new event in the practitioner's mind. This directed the educational project to look at the composition of solutions in contexts that novice practitioners can develop expertise in and to critically appreciate the limits of category knowledge that is taught.

The research also revealed that experienced practitioners could use their knowledge when the situation was outside their experience. Their levels of appreciation and ways of working were congruent such that they could actively create knowledge for action in new situations. Thus, they could work when their knowledge was incomplete. They still composed solutions from past events but could break the past events down and identify gaps where they needed to inquire further into the current problem. Their working methods included making enquiries to people around the project both inside and outside their organization. Their confidence that they could find a solution when the problem emerged was evident and important although they might initially not know what the solution was. Thus, the development of an ability to work with incomplete knowledge is important for novice practitioners. Indeed, in this knowledge centric approach, it is taken as the key skill.

However, the results also showed that the industry works on a competitive knowledge model; thus, incomplete knowledge is seen as a weakness. Technical knowledge becomes part of a power struggle within the world being used to demonstrate authority of one's position. Thus, incomplete knowledge is a growing problem in the industry anyway as projects and technology become more complex. Many managers in the research were totally solution-focused and found it very difficult to look more broadly at the contributing issues to problems always expressing the time constraints as a reason not to learn. This was identified as a barrier to knowledge management but was also an attitude that novices needed to, at least, understand and, hopefully, avoid in their development. This jumping to solutions prevents the industry developing more productive practices and novices need to use events, but use them critically.

Critical and reflective working with incomplete knowledge is the key to success for conversion students. What novice practitioners lack is a quantity of past events as well as compositional skills. Thus, the course should provide both of these. The knowledge management research project could help in this education because it had more than 300 events. Compositional skills require inquiry into the situation, problem modelling, lateral thinking, critical appreciation and a critical self-confidence (De Bono, 2004). In knowledge management applications (and education), there has been an almost universal focus on *what is known* as this is the capturable, codifiable and tradable asset as conceptualized by a product view of knowledge (Empson, 2001). The new thinking postulated here is that it is not what a practitioner knows that is important, but how they *handle situations of incomplete knowledge*. This is congruent with a process view of knowledge (Empson, 2001) where knowledge is socially based as part of situated experiences (Lave and Wenger, 1991) and where there is a problem-solving approach to design (De Bono, 2004). Clearly, this handling of incompleteness cannot be encapsulated in the same way as capturing solutions; however, the project has identified a number of ways it can be understood, supported and enhanced so that it can be part of knowledge management and of education.

KNOWLEDGE CENTRIC EDUCATION

The knowledge centric approach involves developing understanding and skills in handling incomplete knowledge and critical reflection on the operation of the industry in order to learn from and improve practice.

Conversion students will have incomplete knowledge for a long time in practice because of their short time in education and their lack of experience. A knowledge centric approach sees the development of being practically adequate with incomplete knowledge as the critical factor for conversion students. Thus, the students need to be 'knowledge-enabled' through models, tools and procedures that work with incomplete knowledge. This approach deviates from conventional approaches, based on detailed technical knowledge, by developing a capability to work with inadequate knowledge. Other students are knowledge inadequate but have no tools to deal with this and so cover up their deficit and fail to learn, a symptom that is endemic within the industry (Boyd and Wild, 1993). This may also have significance for undergraduate education; however, the educational context and risk make conversion education the appropriate place to develop it.

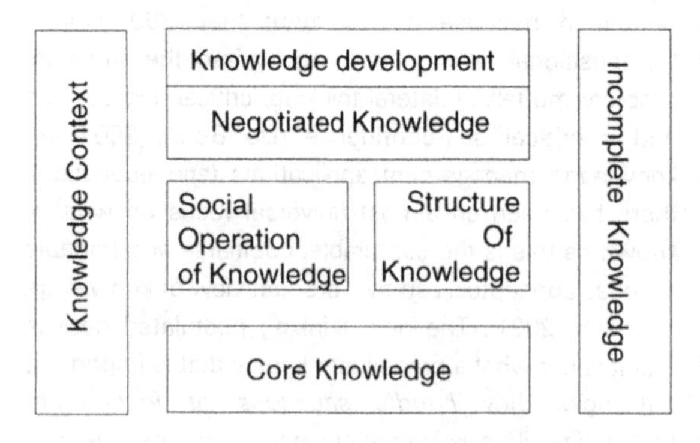

FIGURE 1 Knowledge centric education framework

The advantage is that conversion students are educationally aware and so capable of engaging explicitly in a new process. This acknowledges the graduate character of the individuals as well as establishing a postgraduate level of education. The disadvantage is that the educational time is very short. The delivery of undergraduate technical education in a shorter period of time is not physically possible (although adopted on other conversion courses). In a knowledge centric approach, it is not desirable for the future careers of these people as it reproduces failed cultural models of operation of the industry.

The knowledge centric educational framework is shown in Figure 1. The approach still requires a core technical background as this is the way knowledge is held and operated in the industry and by professions. The convenience of educating through these normal categories is irresistible, however, the problems of it have never been acknowledged. The rest of the model in Figure 1 addresses this.

Schön (1983) presents a very well developed critique of the application of the technical rationality (or category knowledge) view of practice. Category knowledge is problematic as it stops novice practitioners dealing with problems holistically. It directs their thinking to pigeon-hole problems within the categories from which the solution of the category is chosen. Although construction education has involved students undertaking simulation projects, there is an educational barrier set by course structures around category knowledge. In education simulation projects, problems tend to be approached from a category knowledge perspective and, more importantly, assessed through this. In the latter case, this means that assessors use the 'lens' of their category background to evaluate and criticize, which emphasizes students' failing category knowledge rather than composition of solutions. This is often also profession-dependent which emphasizes the fragmentation of the industry. Students are torn between assessment achievement and practice and will chose the assessment perspective. Individually, they will excel at the category knowledge based on their preferences (i.e. learning styles, see Honey and Mumford, 1992) rather than develop solution composition skills that require developing knowledge away from their preferences.

When novice practitioners are in practice, they develop a broader appreciation of solutions; however, this is haphazard and uncritical so that novice practitioners may simply mirror the inappropriate knowledge of experienced practitioners who are themselves reproducing problematic approaches. In our desire to change the industry, we need novice practitioners to learn from practice but not to simply reproduce practice. This requires the rest of the knowledge centric framework into which core knowledge can be placed in order to appreciate critically its significance.

Thus, there is a tension in knowledge centric education between the quantity of core knowledge that individuals require in order to undertake tasks, and the development of skills of thinking, acting and learning to work within the environment that they find themselves. This can only be resolved by the students themselves through the other elements of the model in Figure 1 by developing a self awareness of their own learning approach and current incomplete knowledge, an ability to learn in different situations, a comprehension of the constitution of different knowledge, a knowledge of the codification of technical knowledge (not just the knowledge itself) so that they can access from any stored form and also evaluate any knowledge, a knowledge of the social operation of technical knowledge both to be better able to access technical knowledge as well as to be sensitive to situations when technical knowledge is negotiated or abused. And, finally, they must be aware of how new knowledge (and practice) is developed and sustained, particularly as the expectation is that they are going to be at the forefront of this. This also acknowledges the graduate character of the individuals in conversion education as well as establishing a postgraduate level of educational content.

This model is not just an educational model but also a model for the continuing development of a professional. As experts enter new situations they need to reflect on their incomplete technical knowledge in the context, access and evaluate new knowledge, appreciate the social operation of knowledge in the situation, and effectively negotiate with their knowledge. In the end, the new situation has the potential for knowledge development. Thus, the operation of the model is not linear from the bottom up, but cyclical with different rates and extents of iteration. This is particularly important as all professionals' work changes in their career becoming more managerial rather than technical. It also enables the development of new practice in the industry.

EDUCATIONAL DELIVERY

It was decided that the best educational delivery would be part-time as this allows students to more rapidly appreciate both construction technology and the situated experience (Lave and Wenger, 1991) of the industry such that their work could be used as part of their educational development. To this end, the course is being run in association with a number of consultancy and contracting companies that will be involved in recruiting such entrants and seconding them on the course. Attendance is one full day per week over two full academic years plus a number of residential weekends. The final MSc component of the course is offered on a flexible basis beyond the basic two-year period. The course is supported by both the Chartered Institute of Building (CIOB) and the Royal Institution of Chartered Surveyors (RICS) and the aim is to present candidates for professional membership.

The course leads to an MSc in construction through three interim qualifications – graduate certificate, postgraduate certificate and postgraduate diploma in construction – as shown in Table 3. The qualifications are based on four themes to engage with the knowledge centric approach of the course – core knowledge, knowledge in practice, critical analysis of practice and knowledge development in practice. Given the educational, professional and practice norms, it was not possible to be completely radical in content and delivery. The new approach has to be developed alongside conventional approaches. Each theme has four modules. The key introductory module of the knowledge centric approach – learning in practice – sits within the second of these themes but commences at the start of the course as the students acquire their core knowledge. The knowledge centric theme is taken up explicitly in the two project modules and in the critical evaluation of practice.

LEARNING IN PRACTICE MODULE

The learning in practice module is key and is designed to enable conversion students to develop rapidly an

TABLE 3 Course structure

Theme 1 Core knowledge Graduate certificate (level 6)	Construction Technology (double unit)	Law	Health and safety	Cost and finance
Theme 2 Knowledge in practice Postgraduate certificate (level 7)	Learning in practice	Development policy and practice	Practice Project	Management
Phase 3 Critical analysis of practice Postgraduate diploma (level 7)	Supply chain management and contract administration	Operation strategy Project	Critical evaluation in practice	Design cost and value
Phase 4 Knowledge development in practice MSc (level 7)	Dissertation			

effective working style within the construction industry and to continue to develop throughout their careers. The module approach is based on personal knowledge management techniques. A number of models of personal learning and knowledge management are presented (Kolb, 1984; Honey and Mumford, 1992). These are used initially for the students to analyse themselves to appreciate how they learn effectively and how they can take responsibility for their own learning in new situations. The particular difficulty for conversion students of inadequate knowledge is highlighted for individuals to develop personal approaches to this problem both to cope with situations and to use them for knowledge development. Higher-level knowledge acquisition techniques in complex technical and social situations are explored using cases written from the knowledge management research project.

The development of skills in the composition of critical solutions is approached using examples of professional education that are moving away from a pure knowledge-transfer model to a knowledge-in-practice model for the novice to become an independent learner (Taylor, 1997). This is already the case in people-centred professions such as teaching (Atkinson and Claxton, 2000), social work (Newman and Holzman, 1997) and nursing (Cleverly, 2003). Conventional construction education resides substantively in a technical world where knowledge is separated from practice then reapplied to practice. Thus, advances in construction education from this perspective have focused on getting students to maintain portfolios of skills that they have applied either in work or in simulated projects (Maddocks and Wright, 2004). A different approach was sought and inquiry-based learning (Churchman, 1971; Cleverly, 2003) adopted, as this had the notion of incompleteness handled through finding out embedded in it. In a socially based 'situated experience' (Lave and Wenger 1991), where knowledge is incomplete, such as in construction, then actions are negotiated from: norms of knowledge; social and organizational positions including status; power and authority; perceptions of risk including loss of face as well as technical unknowns; and personal friendships. Thus, an understanding of the sociology of construction is critical to appreciating and learning from a world of practice and this is developed as part of the course.

A wider context of the industry also needs to be presented to enable students to place knowledge in an institutional framework including an understanding of the conflicts, both social and technical, that may be experienced. This provide's an understanding of the constitution of knowledge in practice including a critical review of the position of law, regulation, best practice, standards, etc. The problems of information management, roles and personalities in project teams are introduced for resolution later in the course in the project modules. Of particular importance is the idea that novice practitioners acquire knowledge by observing other people during their work. The idea of transfer of tacit knowledge through socializations (Nonaka and Takeuchi, 1995) involves being fully with someone and experiencing the whole situation and their model practitioner's response to it. Such learning is acknowledged in neuro linguistic programming (NLP) where individuals access the mental states of a model in order to learn (Knight, 1995). Again, the course encourages the novice to be critically aware rather than simply copying behaviour.

Assessment is based on students creating their own individual learning-from-practice portfolio which includes self analysis and social, technical and personal analyses of situations. Students share and develop these analyses in class to give them confidence to admit incomplete knowledge, to gain a wider appreciation of situations from others' perspectives and to learn from others.

CORE KNOWLEDGE

These modules initiate the development of sufficient technical knowledge – in the categories of construction technology, construction accounting, law and regulation – for the students to appreciate situations and to communicate with others. They are initially based on conventional approaches to technical learning, simply acknowledging the technical rationality and knowledge incompleteness problems. There is a difficulty in teaching technical knowledge only because of quantity, as it has been generalized and codified into an explicit rational form. There are great pressures to present the greatest quantity of technical knowledge that is possible. These pressures come from the industry, the students and also the teaching staff. For the industry, there is a fear that inadequate knowledge will lead to mistakes or be used as a weapon to take advantage in project negotiations. For students, there is a feeling of inadequacy and incompetence at not knowing that would be alleviated by more knowledge. For teaching staff, there is a simplicity in the delivery and assessment of this knowledge. Such pressures are understandable and must be recognized, hence the importance of the learning in practice module. It is also acknowledged that technical knowledge can be disputed, although this is seldom recognized in its codified form, and an understanding of this is part of its appropriate application.

KNOWLEDGE IN PRACTICE

The core units stand outside of practice and inform practice. These modules start to explore how knowledge is used in practice. This theme includes the key introductory module learning in practice, but also includes modules on management, which outline the operation of organizations and teams, and on development policy and practice, which explore the context in which buildings are created. The practice option module is designed to explore how knowledge is used holistically and effectively in practice and this is a core competency for practitioners.

CRITICAL ANALYSIS OF PRACTICE

This theme allows students to not just investigate but also to evaluate practice. This bridges an academic and theoretical study with a deeper and critical awareness of practice. Such skills are useful for understanding the idiosyncrasies of construction practice but also for enabling the students to be part of changing practice. The modules are integrative of previous work on the course and of experience from practice. The design, cost and value module considers the complexity of decision making within the industry and addresses the conflicts that are endemic in order to explore solutions in theory and in practice. The supply chain management and contract module uses the current in-vogue perspective on improving the industry to look at the overall operation of practice and in particular exploring how contract is structured and operates within the industry. The operation strategy module investigates and develops what practitioners really do in practice. This again relates to the student's core work competency by developing a mature understanding of role and function in order to more rapidly generate a competent practitioner. The last module, critical evaluation of practice, again takes the work situation but looks at this holistically in relation to the wider project and industry success in particular exploring the way different perspectives are accommodated within the industry.

KNOWLEDGE DEVELOPMENT IN PRACTICE

The final theme involves the dissertation and course input surrounding research methodology and how research can be used to develop practice. This is the culminating achievement of the student and will prepare them for the intellectual rigour and critical confidence required later in their careers.

CONCLUSION

This paper has introduced the theory and practice of a radical development in construction education formed through research into knowledge management in practice. This education – called a knowledge centric approach – is based on the awareness that complete knowledge in a practice situation is not available and so what is required is the development of skills in handling incomplete knowledge. This is the case generally but it is critical for conversion students. This requires a change in thinking away from knowledge as an asset and an appreciation of knowledge in its socially situated setting. Tools and techniques are then taught that enable knowledge inquiry and the development of skills of knowledge negotiation in practice. The challenge of the knowledge centric approach is to create practitioners who are adequate with incomplete knowledge and this skill becomes the root of their continuous professional development, and to bring about an effective change in the industry.

AUTHOR CONTACT DETAILS

David Boyd: School of Property, Construction and Planning, University of Central England, Birmingham, B42 2SU, UK.
Tel: +44 (0) 121 331 5233, fax: +44(0) 121 331 5172
e-mail: david.boyd@uce.ac.uk

REFERENCES

Abbott, A., 1988, *The System of Professions*, Chicago, University of Chicago Press.

ACCE, 1995, *Accreditation Standards for Construction Programs*, San Antonio, Texas, American Council for Construction Education.

Anderson, J., 1992, *An Assessment of Education and Training Needs among Construction Personnel*, Texas, Construction Industry Institute.

Andrews, J. and Derbyshire, A., 1993, *Crossing Boundaries*, London, Construction Industry Council.

Argyris, C. and Schön, D.A., 1974, *Theory in Practice: Increasing Professional Effectiveness*, San Francisco, Jossey-Bass Publishers.

Atkinson, T. and Claxton, G., 2000, *The Intuitive Practitioner: On the Value of not always Knowing what One is Doing*, Buckingham, The Open University Press.

Baron, R. and Byrne, D., 1994, *Social Psychology*, 7th edn, Boston, Allyn and Bacon.

Baumard, P., 1999, *Tacit Knowledge in Organizations*, London, SAGE Publications.

Boyd, D., Egbu, C., Chinyio, E., Xiao, H. and Lee, C., 2004, 'Learning from SME site managers through debriefing', in R. Ellis (ed.), *Proceedings of COBRA 2004 Conference, Leeds, September 2004*, London, RICS Foundation.

Boyd, D. and Pierce, D., 2001, 'Implicit knowledge in construction professional practice', in A. Akintoye (ed.), *17th Annual Conference of the Association of Researchers in Construction Management (ARCOM), Salford, September 2001*, 37–46.

Boyd, D. and Wild, A., 1993, 'Innovation and learning in construction project management', in R. Eastham and M. Skitmore (eds), *9th Annual ARCOM Conference, Oxford, September 1993*, 95–106.

BSI, 2001, *Knowledge Management: A Guide to Good Practice, PAS 2001*, London, British Standards Institution.

Business Roundtable, 1982, *Management Education and Academic Relations: Construction Industry Cost Effectiveness Report A5*, Cincinnati, Construction Users Round Table.

Castells, M., 1996, *Information Age: Economy, Society and Culture: Rise of the Network Society*, Vol. 1, (The Information Age: Economy, Society & Culture), Oxford, Blackwell Publishers.

CEBE, 2003, *Rethinking Construction Education: Halting the Demise of UK Construction*, Salford, Strategic Forum/Centre for Education in the Built Environment.

Churchman, C. West, 1971, *The Design of Inquiring Systems*, New York, Basic Books.

CIB, 1996, *Educating the Professional Team*, WG9, London, Construction Industry Board.

Cleverly, D., 2003, *Implementing Inquiry-Based Learning in Nursing*, London, Routledge.

Dawson, R., 2000, *Developing Knowledge Based Client Relationships*, Oxford, Butterworth-Heinemann.

De Bono, E., 2004, *Edward De Bono's Thinking Course*, London, BBC Books.

Easterby-Smith, M., Thorpe, R. and Lowe, A., 1991, *Management Research: An Introduction*, London, Sage.

Egan, J., 1998, *Rethinking Construction*, London, DTI.

Empson, L., 2001, 'Introduction: knowledge management in professional services firms', in *Human Relations*, 54(7), 811–817.

Franklin, U., 1992, *The Real World of Technology*, Ontario Anonsi.

Gann, D. and Salter, A., 1999, *Interdisciplinary Skills for Built Environment Professionals*, London, The Ove Arup Foundation.

Goleman, D., 1996, *Emotional Intelligence: Why It Can Matter More Than IQ*, London, Bloomsbury.

Honey, P. and Mumford, A., 1992, *Manual of Learning Styles*, Maidenhead, Honey and Mumford Press.

Horvath, J.A., Forsythe, G.B., Bullis, R.C., Sweeney, P.J., Williams, W.M., McNaly, J.A., Watendorf, J.M. and Sternberg, R.J. 1999, 'Experience, knowledge and military leadership' in R.J. Sternberg *et al* (eds), *Tacit Knowledge in Professional Practice: Researcher and Practitioner Perspective*, Mahwah, NJ, Lawrence Erlbaum Associates, 39–57.

Jamous, H. and Peloille, B., 1970, 'Changes in the French university hospital system', in J.A. Jackson (ed.), *Professions and Professionalisation*, Cambridge, Cambridge University Press, 110–152.

Khatri, K. and Ng, H.A., 2000, 'The role of intuition in strategic decision making', in *Human Relations*, 53(1), 57–86.

Knight, S., 1995, *NLP at Work*, London, Nicholas Brealey Publications.

Kolb, D., 1984, *Experiential Learning*, New York, Prentice-Hall.

Larson, M.S., 1977, *The Rise of Professionalism*, Berkeley, CA, University of California Press.

Latham, M., 1994, *Constructing the Team*, London, HMSO.

Laurillard, D., 2001, *Rethinking University Teaching*, London, Routledge Falmer.

Lave, J. and Wenger, E., 1991, *Situated Learning. Legitimate Peripheral Participation*, Cambridge, University of Cambridge Press.

Lenard, D., 2004, 'The new seats of learning', *Construction Manager* February 2004, 11.

Macdonald, K., 1995, *The Sociology of the Professions*, London, Sage.

Maddocks, A. and Wright, I., 2004, 'The RAPID progress file: a tool to facilitate work placements', in *CEBE Transactions*, 1(1), 39–47.

Newman, F. and Holzman, L., 1997, *The End of Knowing: A New Developmental Way of Learning*, New York, Routledge.

Nonaka, I. and Takeuchi, H., 1995, *The Knowledge-Creating company: How Japanese Companies Create the Dynamics of Innovation*, Oxford, Oxford University Press.

Patel, V.L.A., Arocha, J. and Kaufman, D., 1999, 'Expertise and tacit knowledge in medicine' in R.J. Sternberg and J.A. Horvath (eds), *Tacit Knowledge in Professional Practice: Researcher and Practitioner Perspectives*, Mahwah, NJ, Lawrence Erlbaum Associates 75–99.

Payne, J. and Sheehan, T., 2004, *Demystifying Knowledge Management*, London, Construction Excellence.

Pearson, I., 2003, 'Public sector – the last stronghold of professional altruism?', Paper 4 in series *Exploring Professional Values for the 21st Century*, London, RSA.

Pearson, M. and Smith, D., 1985, 'Debriefing in experience-based learning' in D. Boud *et al* (eds), *Reflection: Turning Experience into Learning*, London, Kogan Page, 69–84.

Polanyi, M., 1966, *The Tacit Dimension*, London, Routledge and Kegan Paul.

Schön, D., 1983, *The Reflective Practitioner*, New York, Basic Books.

Strategic Forum, 2003, *Accelerating Change Report 2002*, London, DTI.

Taylor, I., 1997, *Developing Learning in Professional Education*, Buckingham, Society for Research in Higher Education and Open University Press.

Weick, K., 1995, *Sensemaking in Organisations*, London, Sage.

Wenger, E., 1998, *Communities of Practice*, New York, Cambridge University Press.

NOTES FOR CONTRIBUTORS

Architectural Engineering and Design Management

1. SUBMISSION

Authors should submit one copy of the printed manuscript plus one copy on disk or via email. Authors should ensure that the disk corresponds to the final revised version of the manuscript hard copy exactly. Digital files should be named as follows: author's surname - keyword - date of submission e.g. jones collaborative 251204. All authors are asked to submit full contact details for three potential reviewers of their manuscript. Manuscripts should be submitted to:

Professor Dino Bouchlaghem, Department of Civil and Building Engineering, Loughborough University, Leicestershire, LE11 3TU, UK. Email: n.m.bouchlaghem@lboro.ac.uk

Authors should keep a copy of the articles and illustrations.

While the Editors, Referees and Publishers will take all possible care of material submitted to the Journal, they cannot be held responsible for the loss of or damage to any material in their possession.

All articles will be peer-reviewed before acceptance. The final decision on acceptance will be made by the appropriate editor.

2. LANGUAGE AND STYLE

Articles should be in English and should be written and arranged in a style that is succinct and easy for readers to understand. Authors who are unable to submit their articles in English should contact the Editors so that any alternatives may be considered. Illustrations should be used to aid the clarity of the article; do not include several versions of similar illustrations, or closely-related diagrams, unless each is making a distinct point.

3. MANUSCRIPT PREPARATION AND LAYOUT

The manuscript should be printed in double-spacing on one side of the paper (A4/letter) only with a 5-cm / 2-inch wide margin on the left hand side. The pages should be numbered consecutively. Headings and subheadings should be used so that the paper is easy to follow.

The first page of the manuscript should contain the full title of the article, the author(s) names without qualifications or titles, and the affiliations and full address of each author. The precise postal address, telephone and fax numbers and email address of the author to whom correspondence should be addressed should also be included.

The second page should contain an abstract of the article and a key word list (5-10 words). The abstract should be no more than 200 words long and should précis the article, giving a clear indication of its conclusions. The maximum length for the entire manuscript is 7000 words.

Tables and Schema

Each of these should be on a separate sheet, at the end of the manuscript, clearly labelled with the name of the main author. Authors should aim to present table data as succinctly as possible and tables should not duplicate data that are available elsewhere in the article.

Symbols, Abbreviations and Conventions

Please use SI (Systeme Internationale) units. Whenever an acronym or abbreviation is used, ensure that it is spelled out in full the first time it appears. Please indicate in the margin any unusual symbols such as Greek letters that are used in the article.

References and Notes

References should be presented in 'author/date' style in the text and collected in alphabetical order at the end of the article. All references in the reference list should appear in the text. Each reference must include full details of the work referred to, including paper or chapter titles and opening and closing page numbers.

JOURNALS

Smith, D.W., Davis, L.M. and Price, B.N., 2004, 'Integrated Decision-Making in Construction', in *Architectural Engineering and Design Management,* **1**(1), 64–72.

BOOKS

Price, B.N., 1993, 'Integrated Decision-Making in Construction', in L.M. Davis, (ed.), *Integrated Thinking*, London, James & James (Science Publishers) Ltd, 84–92.

Smith, D.W. and Davis, L.M., 1993 *Integrated Thinking*, London, James & James (Science Publishers) Ltd.

PROCEEDINGS

Smith, D.W., Davis L.M. and Price, B.N., 1999, 'Integrated Decision-Making in Construction', in *Proceedings of the 8th Annual Meeting of the Society for Architectural Engineering*, London, James & James (Science Publishers) Ltd.

Notes – which should be kept to a minimum – will appear as endnotes. Indicate endnotes with a superscript number in the text, and include the text at the end of the article. Do not use the footnote/endnote commands in word processing software for either references or notes.

Illustrations

Illustrations should be included at the end of the manuscript. Photographs and line drawings should be referred to in the text as Figure 1, Figure 2. etc. Each illustration requires a caption.

Illustrations should be numbered in the order in which they appear. They should be submitted in a form ready for reproduction – no redrawing or re-lettering will be carried out by the Publishers. Each illustration should be clearly market with the figure number, the name of the main author, and the orientation if it is needed.

Images can be supplied either as hard-copy originals or digitally. Hard-copy photographs should be prints with good definition. Graphs and diagrams should be laser-printed on good quality white paper. Digital images must be supplied as separate tiff or jpeg files, and not embedded in the text file. Images must be at least 300 dpi at 140mm wide. Note that figures and graphs must be comprehensible in black-and-white – use patterns, not colours, to differentiate sections. If colour is essential, in most cases the additional cost for including colour will be borne by the author.

Most line drawings will be reduced in size for publication, usually by about 50%. Please bear this in mind and try to ensure that all lines are at least 0.25mm thick and lettering is in upper and lower case with capitals at least 4mm high.

4. PROOFS AND OFFPRINTS

The corresponding author will receive proofs for correction; these should be returned to Earthscan within 48 hours of receipt. The corresponding author will be sent a pdf of the published article. If printed offprints are required, these must be ordered prior to publication; an order form will be provided for this purpose.

5. COPYRIGHT

Submission of an article to the journal is taken to imply that it represents original work, not under consideration for publication elsewhere. Authors will be asked to transfer the copyright of their articles to the publishers. Copyright covers the distribution of the material in all forms including but not limited to reprints, photographic reproductions and microfilm. It is the responsibility of the author(s) of each article to collect any permissions and acknowledgements necessary for the article to be published prior to submission to the Journal.

9781844073306